CHEMISTRY

(Large Size & Large Print Edition)

Chris McMullen
Northwestern State University of Louisiana

Understand Basic Chemistry Concepts (Large Size & Large Print Edition): The Periodic Table, Chemical Bonds, Naming Compounds, Balancing Equations, and More.

Third edition (February, 2013)

Astro Nutz

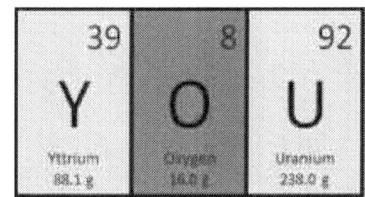 

ISBN-10: 1479201251.
ISBN-13: 978-1479201259.

Chris McMullen

Northwestern State University of Louisiana

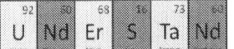 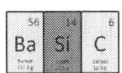 CHEMISTRY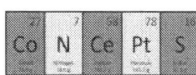

Contents

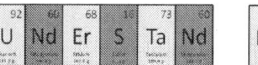 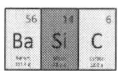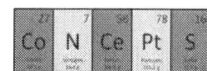
 Introduction

 is book focuses on fundamental chemistry concepts, such as understanding the periodic table of the elements and how chemical bonds are formed. No prior knowledge of chemistry is assumed. The mathematical component involves only basic arithmetic.

The coverage of chemistry is not comprehensive, but does cover a variety of main concepts. The content is much more conceptual than mathematical. It is geared toward helping anyone – student or not – to understand the main ideas of chemistry.

The verbal side of your brain may appreciate that this book is decorated with chemical words. A chemical word is a word that can be made by only using symbols from the periodic table. For example, the following sentence is made entirely of chemical words:

May you enjoy your chemical journey! ☺

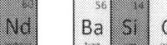  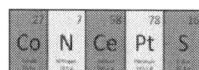

1 The Periodic Table

1.1 Science and Chemistry

hat is chemistry? That's one of those seemingly simple questions that can frustrate students if it appears as a short answer question on a final exam. Nevertheless, if you study chemistry – whether leisurely or seriously – it might be helpful to have some idea of just what chemistry is.

Before we answer this question, let's consider another one: What is science? Science is the knowledge of nature that is obtained systematically through observation, analysis, and experimentation.

- Science is based on experiment. Every principle of science can be tested, and only remains valid as long as it is confirmed by experiment.
- Science is systematic. Scientists follow the scientific method.

The scientific method involves the following steps: Observation (preliminary measurements), hypothesis (an educated guess for how to explain the data), experimentation (more measurements), interpretation (data analysis), theory (developing a model to explain the pattern), additional experimentation (testing the theory), and predictions (based on a refined model). Predictability makes science extremely useful.

Eventually, the theory may become an established law. However, even laws only remain valid as long as they continue to be supported by experiment.

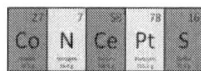

Every branch of science can be classified as a physical science or a life science. A life science involves the study of living things. Examples include biology, anatomy, and physiology. A physical science involves the study of nonliving things. Examples include chemistry, astronomy, and physics.

Back to our original question: What is chemistry? Chemistry is the science that involves the study of different types of atoms, their properties, and the types of bonds that they form. Chemistry is often called the central science because all of the applied sciences – like geology and biology – involve chemistry.

Chemistry deals with chemical, as opposed to physical, change. When a substance undergoes a physical change, the composition remains the same, but the motion of the molecules or the separation between molecules changes. An example of a physical change is the melting of ice into water. Both ice and water consist of H_2O molecules – i.e. molecules that have two hydrogen (**H**) atoms and one oxygen (**O**) atom joined together.

When a substance undergoes a chemical change, a new substance is formed with a different composition. The new molecules are different from the old molecules. An example of a chemical change is the rusting of iron. Separate **Fe** (iron), O_2 (oxygen gas), and H_2O (water) molecules combine to form $Fe(OH)_3$ molecules.

1.2 Classifying Matter

 I things in the universe are either forms of matter or radiation. Your clothes, this book, furniture, homes, cars, your body, and even the air that you breathe are forms of matter. Anything that has nonzero rest-mass is a form of matter.

Radiation has zero rest-mass. Light is a form of radiation. Light is made up of particles called photons. A photon doesn't have rest-mass (but you also won't ever find a photon at rest – photons are the fastest-moving

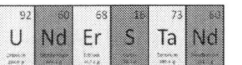 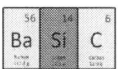 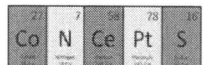

objects in the universe). Photons do have relativistic mass, in accordance with Einstein's famous equation ($E = mc^2$) because photons carry energy.

> ## When an aluminum rod is heated, does it undergo a physical or chemical change?
>
> **Physical**: The molecules move faster on average. The separation between molecules increases, causing the rod to expand. But the molecules still consist of **Al** atoms.
>
> ## When wood burns, is the change physical or chemical?
>
> **Chemical**: New substances are formed, with a different molecular make-up than wood and oxygen.
>
>

Matter comes in many different forms. All forms of matter can be classified as either a pure substance or a mixture. A pure substance has a definite composition. For example, water (H_2O) always has two parts hydrogen (**H**) for each part oxygen (**O**). The ingredients of a mixture do not always come in the same proportions. Air is an example of a mixture. Air has approximately 78% nitrogen (**N**), 21% oxygen (**O**), and 1% of several other components, but the exact composition of air can vary because all of the ingredients of air do not form a specific bond – they are instead separate entities mixed together. The composition of the air varies, as it can easily be polluted or filtered.

There are two kinds of pure substances – elements and compounds. A chunk of pure gold (Au) would consist of a single type of atom, where every single atom has exactly 79 protons, 79 electrons, and approximately 118 neutrons. A chunk of gold would consist of roughly one septillion (that's 1,000,000,000,000,000,000,000,000) such atoms. (If you're curious about the origin of this number, wait until we explore Avogadro's number in Sec. 1.4.) If virtually every atom in the chunk is a gold atom, then the chunk would be a pure substance. Elements are the different types of atoms found on the periodic table (see Sec. 1.3). A substance that consists of a single type of element is one type of a pure substance.

The other type of a pure substance is a compound, which consists of two or more elements combined together in a definite proportion. For example, carbon dioxide (CO_2) consists of molecules with two oxygen (O) atoms bound to a single carbon (C) atom. A compound has a chemical formula – like H_2O or CO_2.

A substance that is not pure (i.e. either an element or a compound) is a mixture. The composition of a mixture is variable – i.e. the ingredients do not always come in the same proportions. There are two types of mixtures – homogeneous and heterogeneous. A homogeneous mixture is essentially a single substance in that it is approximately uniform throughout. A homogeneous mixture is also called a solution. For example, blood is a homogeneous mixture: It is pretty much the same throughout, but doesn't have a definite composition. A heterogeneous mixture consists of two or more substances mixed together non-uniformly. For example, wood is non-uniform. It is visibly grainy.

Following are a few rules for determining how to classify a substance:

- A substance is an element if you can find it on the periodic table. For example, iron is an element with the symbol Fe (atomic number 26). See Sec. 1.3.

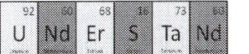 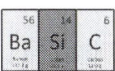  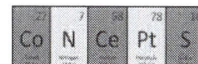

If there is a chemical formula for a substance, it is a compound. For example, sodium oxide is a compound with the formula Na_2O.

A substance that is approximately uniform in all directions is a homogeneous mixture – also called a solution. For example, tea is a solution.

A heterogeneous mixture has some non-uniformity. For example, a rock is a mixture of different minerals. You can see the grains in the rock.

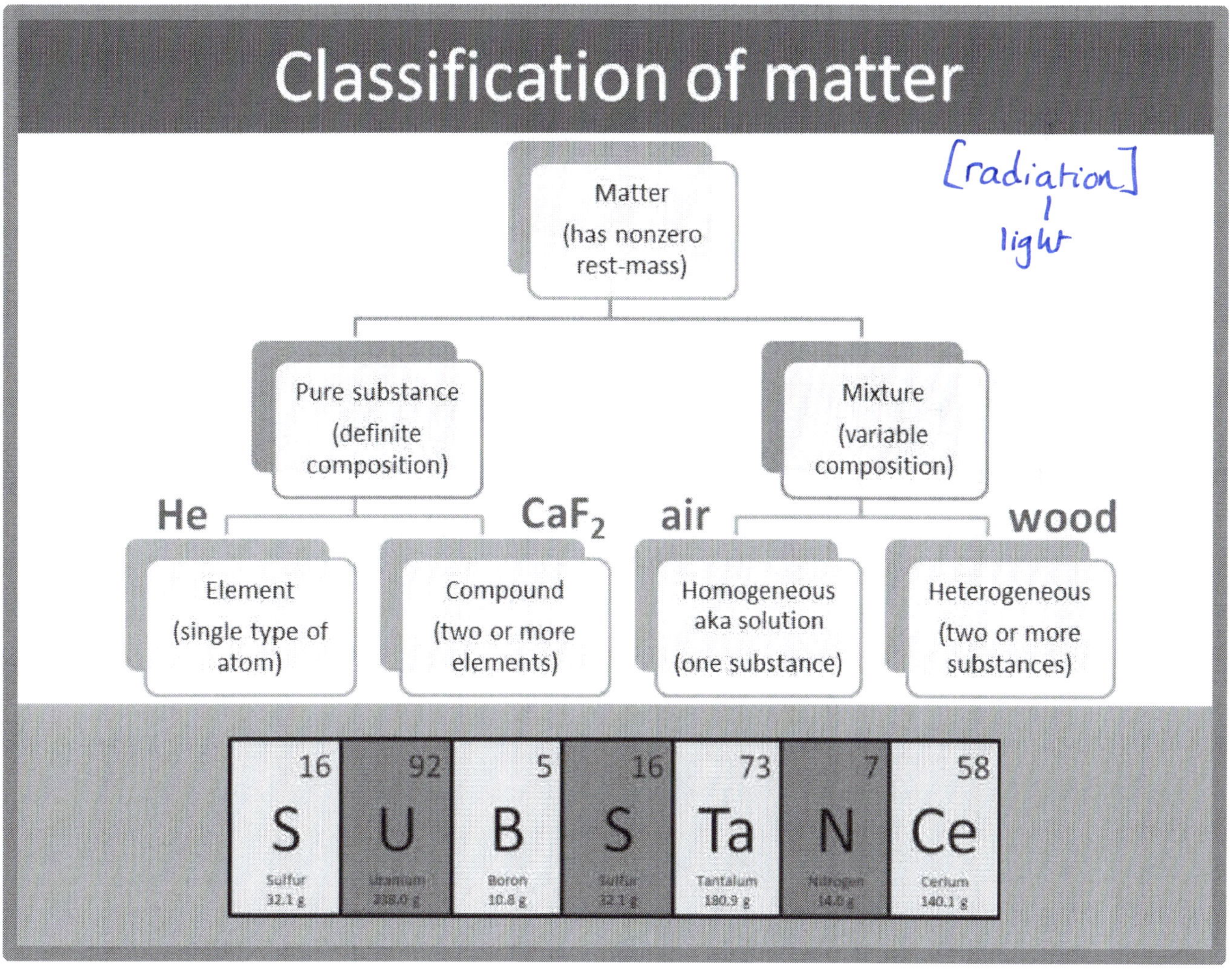

Classification of matter

Matter
(has nonzero rest-mass)

[radiation]
|
light

Pure substance
(definite composition)

Mixture
(variable composition)

He   CaF₂   air   wood

Element
(single type of atom)

Compound
(two or more elements)

Homogeneous
aka solution
(one substance)

Heterogeneous
(two or more substances)

16	92	5	16	73	7	58
S	U	B	S	Ta	N	Ce
Sulfur 32.1 g	Uranium 238.0 g	Boron 10.8 g	Sulfur 32.1 g	Tantalum 180.9 g	Nitrogen 14.0 g	Cerium 140.1 g

State whether each substance below is an element, compound, solution, or heterogeneous mixture:
gold, water, concrete, neon, seawater, carbon monoxide, iron, zinc sulfide, granite, sweet tea

Elements: gold (**Au**), neon (**Ne**), iron (**Fe**). You can find these on the periodic table.

Compounds: water (**H₂O**), carbon monoxide (**CO**), zinc sulfide (**ZnS**). These are definite proportions of two elements.

Solutions: seawater, sweet tea. These are uniform.

Mixtures: concrete, granite. These are non-uniform.

1.3 The Elements

 oms consist of protons, neutrons, and electrons. Protons and neutrons reside in a nucleus at the center of the atom, while electrons surround the nucleus. The negatively charged electrons are attracted to the positively charged protons. Neutrons add mass and stability to the nucleus without affecting the charge.

Positive Protons

Neutral Neutrons

Negative Electrons —odd ones out, in shell not neuclecus.

There are about 100 different types of atoms, which are called elements. They form a pattern which we call the periodic table. Each element has a symbol with one or two letters. For example, **Sn** is the symbol for the element tin.

The elements are numbered according to their atomic number. Atomic number equals the number of protons in the nucleus of the atom. Atomic number starts with hydrogen (**H**) being 1 and increases by 1 going to the right and down (just as you read a book left-to-right across each line, then top-to-bottom down the page). The atomic number of helium (**He**) is 2, lithium (**Li**) is 3, beryllium (**Be**) is 4, boron (**B**) is 5, and so on. This means that a **H** nucleus has 1 proton, a **He** nucleus has 2 protons, a **Li** nucleus has 3 protons, a **Be** nucleus has 4 protons, a **B** nucleus has 5 protons, etc.

The elements are organized by their electron structure. We will explore electron structure in Chapter 2. In the meantime, what you need to know is that the elements in the same column have similar electron structure. This is very important in chemistry because electron structure is the chief factor in understanding the chemical behavior of an element.

It will be helpful to consider the periodic table figures and refer to them frequently as we discuss them.

The periodic table consists of rows and columns. The rows of the periodic table are called periods. For example, lithium (**Li**), beryllium (**Be**), boron (**B**), carbon (**C**), nitrogen (**N**), oxygen (**O**), fluorine (**F**), and neon (**Ne**) are the elements of period 2 of the periodic table. There are 7 periods on the periodic table. The elements that appear to be in rows 8 and 9 are actually part of periods 6 and 7. You can see this if you look at the atomic numbers (the number of protons in the nucleus) in the top right corner of each element's cell (rectangle) on the periodic table.

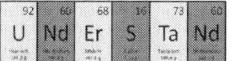 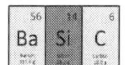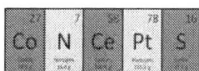
PERIODIC

Atomic Number

Mass per mole

1 **H** hydrogen 1.0 g								
3 **Li** lithium 6.9 g	**4** **Be** beryllium 9.0 g							
11 **Na** sodium 23.0 g	**12** **Mg** magnesium 24.3 g							
19 **K** potassium 39.1 g	**20** **Ca** calcium 40.1 g	**21** **Sc** scandium 45.0 g	**22** **Ti** titanium 47.9 g	**23** **V** vanadium 50.9 g	**24** **Cr** chromium 52.0 g	**25** **Mn** manganese 54.9 g	**26** **Fe** iron 55.8 g	**27** **Co** cobalt 58.9 g
37 **Rb** rubidium 85.5 g	**38** **Sr** strontium 87.6 g	**39** **Y** yttrium 88.9 g	**40** **Zr** zirconium 91.2 g	**41** **Nb** niobium 92.9 g	**42** **Mo** molybdenum 95.9 g	**43** **Tc** technetium 97.9 g	**44** **Ru** ruthenium 101.1 g	**45** **Rh** rhodium 102.9 g
55 **Cs** cesium 132.9 g	**56** **Ba** barium 137.3 g	**57** **La** lanthanum 138.9 g	**72** **Hf** hafnium 178.5 g	**73** **Ta** tantalum 180.9 g	**74** **W** tungsten 183.8 g	**75** **Re** rhenium 186.2 g	**76** **Os** osmium 190.2 g	**77** **Ir** iridium 192.2 g
87 **Fr** francium 223.0 g	**88** **Ra** radium 226.0 g	**89** **Ac** actinium 227.0 g	**104** **Rf** Rutherfordium 261.1 g	**105** **Db** dubnium 262.1 g	**106** **Sg** seaborgium 263.1 g	**107** **Bh** bohrium 262.2 g	**108** **Hs** hassium 265 g	**109** **Mt** meitnerium 266 g

58 **Ce** cerium 140.1 g	**59** **Pr** praseodymium 140.9 g	**60** **Nd** neodymium 144.2 g	**61** **Pm** promethium 144.9 g	**62** **Sm** samarium 150.4 g
90 **Th** thorium 232.0 g	**91** **Pa** protactinium 231.0 g	**92** **U** uranium 238.0 g	**93** **Np** neptunium 237.0 g	**94** **Pu** plutonium 244.1 g

91 **Pa**	**32** **Ge**

TABLE

Group Number

8
2
He
helium
4.0 g

3	4	5	6	7

5	6	7	8	9	10
B	C	N	O	F	Ne
boron	carbon	nitrogen	oxygen	fluorine	neon
10.8 g	12.0 g	14.0 g	16.0 g	19.0 g	20.2 g

13	14	15	16	17	18
Al	Si	P	S	Cl	Ar
aluminum	silicon	phosphorus	sulfur	chlorine	argon
27.0 g	28.1 g	31.0 g	32.1 g	35.5 g	39.9 g

28	29	30	31	32	33	34	35	36
Ni	Cu	Zn	Ga	Ge	As	Se	Br	Kr
nickel	copper	zinc	gallium	germanium	arsenic	selenium	bromine	krypton
58.7 g	63.5 g	65.4 g	69.7 g	72.6 g	74.9 g	79.0 g	79.9 g	83.8 g

46	47	48	49	50	51	52	53	54
Pd	Ag	Cd	In	Sn	Sb	Te	I	Xe
palladium	silver	cadmium	indium	tin	antimony	tellurium	iodine	xenon
106.4 g	107.9 g	112.4 g	114.8 g	118.7 g	121.8 g	127.6 g	126.9 g	131.3 g

78	79	80	81	82	83	84	85	86
Pt	Au	Hg	Tl	Pb	Bi	Po	At	Rn
platinum	gold	mercury	thallium	lead	bismuth	polonium	astatine	radon
195.1 g	197.0 g	200.6 g	204.4 g	207.2 g	209.0 g	209.0 g	210.0 g	222.0 g

110	111	112	113	114	115	116	117	118
Ds	Rg	Cn						
darmstadtium	roentgenium	copernicium						
269g	272 g	277 g						

63	64	65	66	67	68	69	70	71
Eu	Gd	Tb	Dy	Ho	Er	Tm	Yb	Lu
europium	gadolinium	terbium	dysprosium	holmium	erbium	thulium	ytterbium	lutetium
152.0 g	157.3 g	158.9 g	162.5 g	164.9 g	167.3 g	168.9 g	173.0 g	175.0 g

95	96	97	98	99	100	101	102	103
Am	Cm	Bk	Cf	Es	Fm	Md	No	Lr
americium	curium	berkelium	californium	einsteinium	fermium	mendelevium	nobelium	lawrencium
243.1 g	247.1 g	247.1 g	251.1 g	252.1 g	257.1 g	258.1 g	259.1 g	262.1 g

PERIODIC TABLE

1 H hydrogen 1.0 g																	2 He helium 4.0 g
3 Li lithium 6.9 g	4 Be beryllium 9.0 g											5 B boron 10.8 g	6 C carbon 12.0 g	7 N nitrogen 14.0 g	8 O oxygen 16.0 g	9 F fluorine 19.0 g	10 Ne neon 20.2 g
11 Na sodium 23.0 g	12 Mg magnesium 24.3 g											13 Al aluminum 27.0 g	14 Si silicon 28.1 g	15 P phosphorus 31.0 g	16 S sulfur 32.1 g	17 Cl chlorine 35.5 g	18 Ar argon 39.9 g
19 K potassium 39.1 g	20 Ca calcium 40.1 g	21 Sc scandium 45.0 g	22 Ti titanium 47.9 g	23 V vanadium 50.9 g	24 Cr chromium 52.0 g	25 Mn manganese 54.9 g	26 Fe iron 55.8 g	27 Co cobalt 58.9 g	28 Ni nickel 58.7 g	29 Cu copper 63.5 g	30 Zn zinc 65.4 g	31 Ga gallium 69.7 g	32 Ge germanium 72.6 g	33 As arsenic 74.9 g	34 Se selenium 79.0 g	35 Br bromine 79.9 g	36 Kr krypton 83.8 g
37 Rb rubidium 85.5 g	38 Sr strontium 87.6 g	39 Y yttrium 88.9 g	40 Zr zirconium 91.2 g	41 Nb niobium 92.9 g	42 Mo molybdenum 95.9 g	43 Tc technetium 97.9 g	44 Ru ruthenium 101.1 g	45 Rh rhodium 102.9 g	46 Pd palladium 106.4 g	47 Ag silver 107.9 g	48 Cd cadmium 112.4 g	49 In indium 114.8 g	50 Sn tin 118.7 g	51 Sb antimony 121.8 g	52 Te tellurium 127.6 g	53 I iodine 126.9 g	54 Xe xenon 131.3 g
55 Cs cesium 132.9 g	56 Ba barium 137.3 g	57 La lanthanum 138.9 g	72 Hf hafnium 178.5 g	73 Ta tantalum 180.9 g	74 W tungsten 183.8 g	75 Re rhenium 186.2 g	76 Os osmium 190.2 g	77 Ir iridium 192.2 g	78 Pt platinum 195.1 g	79 Au gold 197.0 g	80 Hg mercury 200.6 g	81 Tl thallium 204.4 g	82 Pb lead 207.2 g	83 Bi bismuth 209.0 g	84 Po polonium 209.0 g	85 At astatine 210.0 g	86 Rn radon 222.0 g
87 Fr francium 223.0 g	88 Ra radium 226.0 g	89 Ac actinium 227.0 g	104 Rf rutherfordium 261.1 g	105 Db dubnium 262.1 g	106 Sg seaborgium 263.1 g	107 Bh bohrium 262.2 g	108 Hs hassium 265 g	109 Mt meitnerium 266 g	110 Ds darmstadtium 269 g	111 Rg roentgenium 272 g	112 Cn copernicium 277 g	113	114	115	116	117	118

58 Ce cerium 140.1 g	59 Pr praseodymium 140.9 g	60 Nd neodymium 144.2 g	61 Pm promethium 144.9 g	62 Sm samarium 150.4 g	63 Eu europium 152.0 g	64 Gd gadolinium 157.3 g	65 Tb terbium 158.9 g	66 Dy dysprosium 162.5 g	67 Ho holmium 164.9 g	68 Er erbium 167.3 g	69 Tm thulium 168.9 g	70 Yb ytterbium 173.0 g	71 Lu lutetium 175.0 g
90 Th thorium 232.0 g	91 Pa protactinium 231.0 g	92 U uranium 238.0 g	93 Np neptunium 237.0 g	94 Pu plutonium 244.1 g	95 Am americium 243.1 g	96 Cm curium 247.1 g	97 Bk berkelium 247.1 g	98 Cf californium 251.1 g	99 Es einsteinium 252.1 g	100 Fm fermium 257.1 g	101 Md mendelevium 258.1 g	102 No nobelium 259.1 g	103 Lr lawrencium 262.1 g

 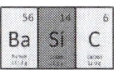
The columns of the periodic table are called groups. For example, fluorine (**F**), chlorine (**Cl**), bromine (**Br**), iodine (**I**), and astatine (**At**) are the elements of group 7 of the periodic table. There are a couple of common, but different, methods for counting the groups. The previous periodic tables show the group number at the top of the column for groups 1-8 in one of the popular numbering schemes. This convention has the conceptual advantage that the group number equals the number of <u>valence electrons</u> for most of the elements in these groups (see Chapter 2).

Elements in the same group (column) exhibit similar chemical behavior. As we will explore in Chapter 2, this is because elements in the same group have similar electron structure. The reason that helium (**He**) appears way over to the right is that helium has similar electron structure and chemical behavior to the elements underneath it. Similarly, this is why there is a gap between beryllium (**Be**) toward the left in group 2 and boron (**B**) further right in group 3.

Elements that exhibit similar chemical behavior are called families. The leftmost group consists of hydrogen (**H**) and the alkali metals (group 1). The alkali metals include lithium (**Li**), sodium (**Na**), potassium (**K**), and so on. Although hydrogen is in the same group, it is not one of the alkali metals. Hydrogen, a nonmetal, does not behave chemically like the alkali metals. As we will see in Chapter 2, hydrogen can gain or lose one electron to fill its outer shell, whereas the other elements in group 1 would need to gain several electrons, but only need to lose one electron, to fill their outer shells. The alkali metals all form bonds with nonmetals by losing one electron.

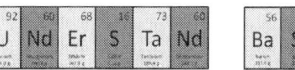 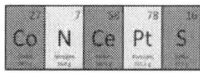
Families on the periodic table

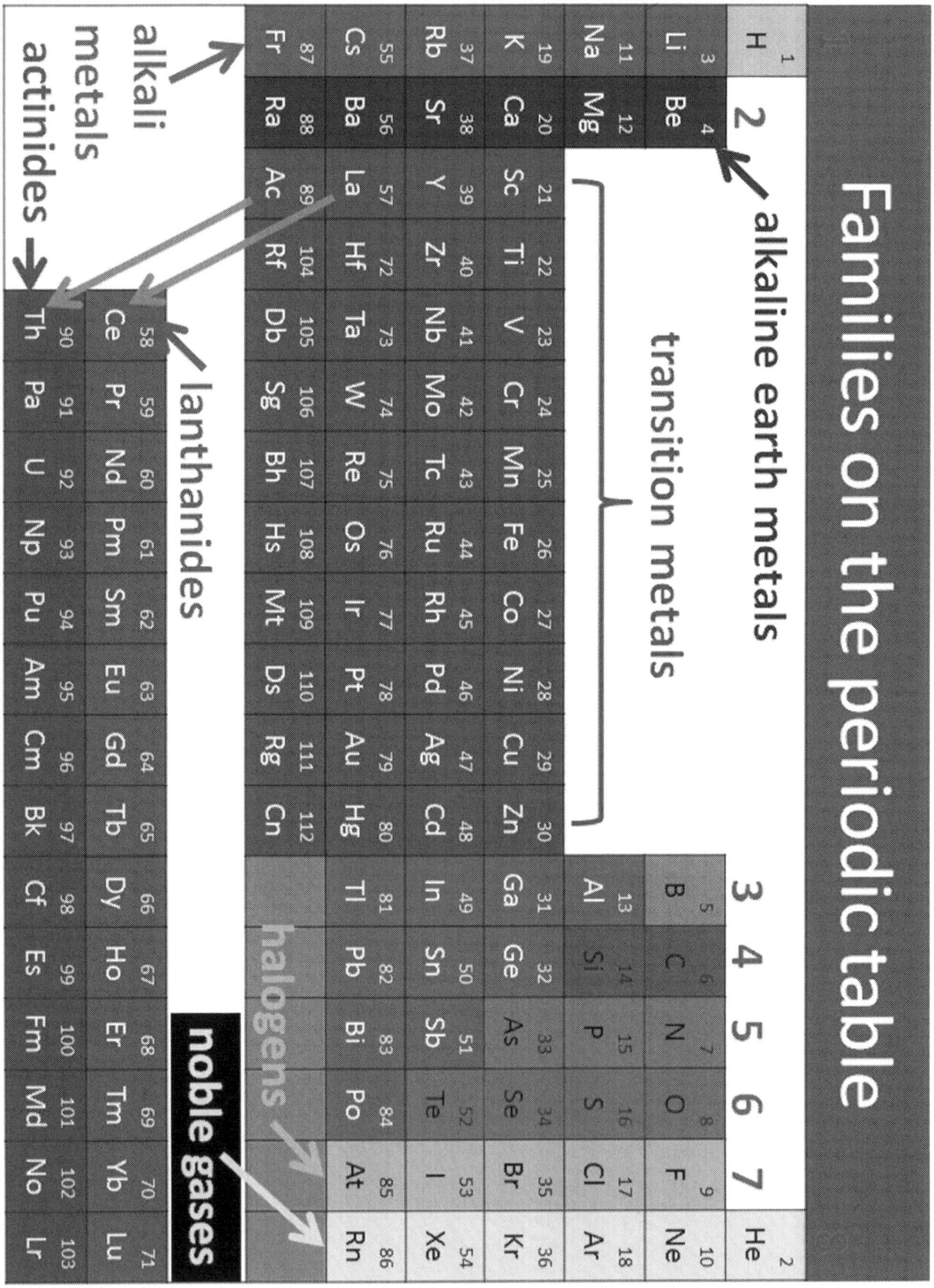

alkali metals

actinides

alkaline earth metals

transition metals

lanthanides

halogens

noble gases

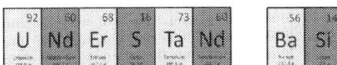 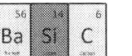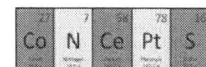
The group second to the left is called the alkaline earth metals (group 2). The alkaline earth metals include beryllium (**Be**), magnesium (**Mg**), calcium (**Ca**), etc. The alkaline earth metals all lose two electrons when forming bonds.

The rightmost group consists of noble gases (group 8), which are also called inert gases. The noble gases are helium (**He**), neon (**Ne**), argon (**Ar**), and so on. The noble gases have filled outer shells and generally do not bond with other atoms.

The second-to-right group consists of halogens (group 7). The halogens include fluorine (**F**), chlorine (**Cl**), bromine (**Br**), etc. The halogens form bonds with metals by gaining one electron and bond with other nonmetals by sharing electrons with them.

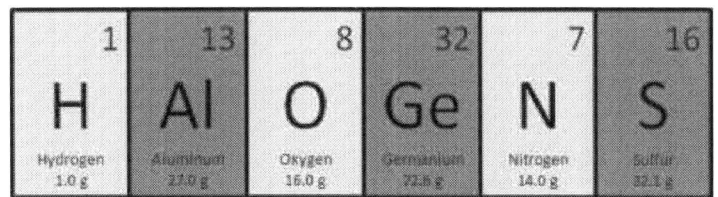

The transition metals are a huge block of similarly behaving elements between groups 2 and 3. The transition metals include scandium (**Sc**) thru zinc (**Zn**) and the elements below them. Shown below the periodic table are parts of periods 6 and 7. The lanthanides include lanthanum (**La**) and the elements cerium (**Ce**) thru lutetium (**Lu**). The actinides consist of actinium (**Ac**) and thorium (**Th**) thru lawrencium (**Lr**). These series fit into the block of transition metals and account for the gaps between atomic numbers 57 and 72 and between 89 and 104 on the periodic table.

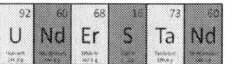 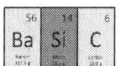 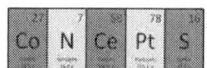

At standard temperature and pressure (STP), the majority of elements are solids. Ironically, "standard" temperature and pressure are not so standard (there are actually a few different common conventions), but generally correspond to room temperature (68°F to 77°F) and atmospheric pressure (1 atm). Only two elements are liquid at STP – mercury (**Hg**) and bromine (**Br**). The elements that are gases at STP include hydrogen (**H**), nitrogen (**N**), oxygen (**O**), fluorine (**F**), chlorine (**Cl**), and the noble gases in group 8. This is illustrated in the figure on the next page.

What is the third element of period 2?
Boron (B).
What is the second element of group 5?
Phosphorus (P).
Which element is in period 3, group 4?
Silicon (Si).
What is the first gas of period 2?
Nitrogen (N).
What is the first solid of group 6?
Sulfur (S).

Solids, liquids, and gases

Legend:
- solid at STP
- liquid at STP
- gas at STP

STP = standard temperature and pressure

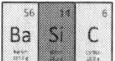 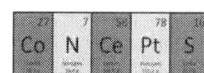

1.4 Periodic Trends

st of the elements are metals, like aluminum (**Al**), zinc (**Zn**), and lead (**Pb**). Other elements are nonmetals, such as sulfur (**S**), oxygen (**O**), and chlorine (**Cl**). A few elements, like silicon (**Si**) and Germanium (**Ge**), are considered to be semimetals, which are called metalloids. The zigzag on the following periodic table divides metals on the left from nonmetals on the right. Note that hydrogen, which is on the left, is an exception to this rule.

Metals and nonmetals have different physical and chemical properties, and exhibit much different chemical behavior (as we will see in Chapter 2). Metals tend to have a shiny luster, metals are opaque (i.e. light does not pass through them), metals are malleable (i.e. they are easily deformed without breaking), and metals tend to be good conductors of both electricity and heat. Nonmetals lack a metallic luster, tend to allow light to pass through, tend to break if you try to deform them (in the solid state), and tend to be poor conductors of electricity and heat.

The reason for the distinction between metals and nonmetals has to do with electron structure, which we will explore in Chapter 2. The main idea is that metals, on the left, tend to lose electrons in order to fill their outer shells, whereas nonmetals, on the right, tend to gain electrons in order to fill their outer shells.

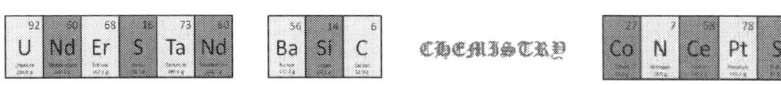

The division between metals and nonmetals

metals

metalloids

nonmetals

1																	2
1 H	2											3	4	5	6	7	2 He
3 Li	4 Be											5 B	6 C	7 N	8 O	9 F	10 Ne
11 Na	12 Mg											13 Al	14 Si	15 P	16 S	17 Cl	18 Ar
19 K	20 Ca	21 Sc	22 Ti	23 V	24 Cr	25 Mn	26 Fe	27 Co	28 Ni	29 Cu	30 Zn	31 Ga	32 Ge	33 As	34 Se	35 Br	36 Kr
37 Rb	38 Sr	39 Y	40 Zr	41 Nb	42 Mo	43 Tc	44 Ru	45 Rh	46 Pd	47 Ag	48 Cd	49 In	50 Sn	51 Sb	52 Te	53 I	54 Xe
55 Cs	56 Ba	57 La	72 Hf	73 Ta	74 W	75 Re	76 Os	77 Ir	78 Pt	79 Au	80 Hg	81 Tl	82 Pb	83 Bi	84 Po	85 At	86 Rn
87 Fr	88 Ra	89 Ac	104 Rf	105 Db	106 Sg	107 Bh	108 Hs	109 Mt	110 Ds	111 Rg	112 Cn						

58 Ce	59 Pr	60 Nd	61 Pm	62 Sm	63 Eu	64 Gd	65 Tb	66 Dy	67 Ho	68 Er	69 Tm	70 Yb	71 Lu
90 Th	91 Pa	92 U	93 Np	94 Pu	95 Am	96 Cm	97 Bk	98 Cf	99 Es	100 Fm	101 Md	102 No	103 Lr

 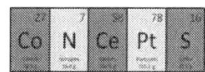

The alkali metals in group 1 all have a single electron in their outer shell. They donate this electron to a nonmetal when forming ionic bonds. It's very easy to lose a single electron, so the alkali metals tend to be very active chemically. Chemical activity (also called electronegativity) is a measure of how readily an element forms bonds.

The alkaline earth metals all need to lose two electrons when forming ionic bonds. They are still highly active metals, but slightly less active than the alkali metals.

There is a general trend in the chemical activities of metals. Going to the left and down on the periodic table, metals tend to become more active chemically. The reason that they become more active going to the left on the periodic table is that elements further to the left don't need to lose as many electrons in order to have filled outer shells.

The noble gases (aka inert gases) in group 8 are very inactive because their outer shells are filled. The other nonmetals become more active going up and to the right on the periodic table. As you go to the right on the periodic table, the nonmetals become more active because they need to gain fewer electrons in order to fill their outer shells. The noble gases are the exception to the trend because they have filled outer shells to begin with and are happy to stay that way. The most active nonmetals are the halogens in group 7.

These trends for the chemical activities of metals and nonmetals are illustrated on the following periodic table. There are a couple of exceptions to the trends for chemical activity and atomic size, but we will base all of the answers to examples and exercises on the trends instead of compiling a list of all of the exceptions. It is much more important to understand the trends than it is to memorize the exceptions.

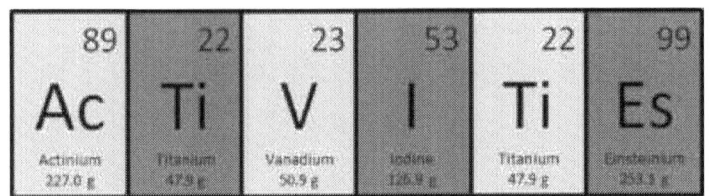

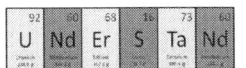 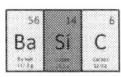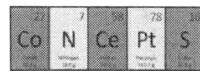
Looking back at the periodic table spread over pages 12-13, the number beneath each element in grams (g) is the molar mass of the element. In Chapter 6, we will learn that the molar mass, in grams per mole, is approximately equal to the number of protons and neutrons in the nucleus of the atom. The molar mass is the mass of one mole of the element, not to be confused with the mass of a single atom.

One mole equals Avogadro's number, which is 6.02×10^{23} to three significant figures in scientific notation. If you don't use scientific notation, the 10^{23} has 23 zeroes: $10^{23} = 100,000,000,000,000,000,000,000$. (We wrote a number with 24 zeroes back in Sec. 1.1 because we had rounded the 6 to 10 in our "estimate.") Avogadro's number is defined as the number of carbon-12 (**C**) atoms needed to make a total mass of 12 g. Carbon-12 is a specific isotope of carbon (we will discuss isotopes in Chapter 6).

Which is more active – phosphorus (P) or chlorine (Cl)?
Chlorine (Cl). These are nonmetals; Cl is further right.
Which is more active – selenium (Se) or oxygen (O)?
Oxygen (O). These are nonmetals; O is above Se.
Which is more active – sulfur (S) or argon (Ar)?
Sulfur (S). Since Ar is a noble gas, it's very inactive.
Which is more active – strontium (Sr) or tin (Sn)?
Strontium (Sr). These are metals; Sr is to the left of Sn.
Which is more active – aluminum (Al) or thallium (Tl)?
Thallium (Tl). These are metals; Tl is below Al.

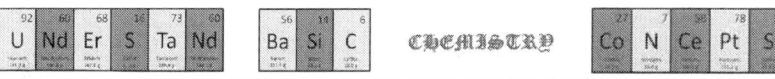
Trends in chemical activity

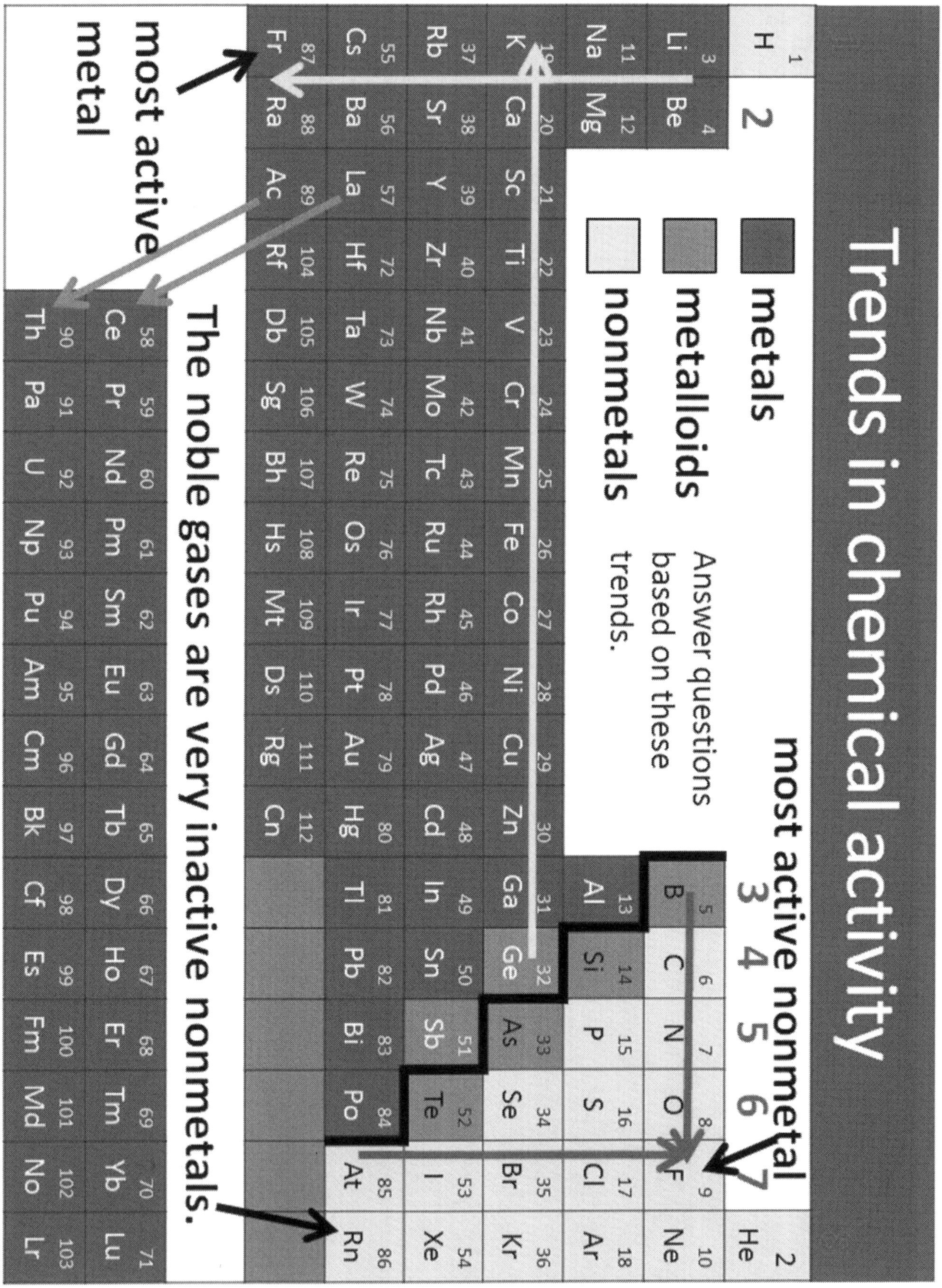

metals

metalloids

nonmetals

Answer questions based on these trends.

most active nonmetal

most active metal

most active metal

The noble gases are very inactive nonmetals.

102	73	22	8	7
No	Ta	Ti	O	N
Nobelium	Tantalum	Titanium	Oxygen	Nitrogen
259.1 g	180.9 g	47.9 g	16.0 g	14.0 g

If you want to know how many helium atoms there are in a container of helium gas, the answer will be an enormous number on the order of 10^{23}. Since it is inconvenient to work with such very large numbers, we introduce the mole. One mole of a substance equals Avogadro's number. If you have 5 moles of helium (**He**) atoms, for example, it's much more convenient to work with 5 moles than it is to work with the number 3.01×10^{24}. We get the number of atoms, 3.01×10^{24}, by multiplying the number of moles times Avogadro's number: $5 \times 6.02 \times 10^{23} = 3.01 \times 10^{24}$.

If you want to know the mass of a single atom of an element, instead of the molar mass, divide the molar mass by Avogadro's number. For example, a single oxygen atom has a mass of $\frac{16.0 \text{ g/mol}}{6.02 \times 10^{23}} = 2.66 \times 10^{-23}$ g. In a chemistry class, you spend a lot of time on such calculations, but the focus of this book is on conceptual understanding, so we will move on now.

27	7	58	78	92	13
Co	N	Ce	Pt	U	Al
Cobalt	Nitrogen	Cerium	Platinum	Uranium	Aluminum
58.9 g	14.0 g	140.1 g	195.1 g	238.0 g	27.0 g

If you have one mole of calcium (**Ca**), the molar mass, which is 40.1 g/mol (you can find this on the periodic tables on pages 12-13), tells you that a scale that measures mass will read 40.1 g. If you have two moles of calcium, the mass will instead be 80.2 g (twice as many moles make twice the mass).

Mass increases with atomic number: The greater the atomic number, the larger the molar mass (and therefore the larger the mass of the atom). Atomic mass increases going to the right and down the periodic table. If you want to compare the atomic mass of two elements, simply look at their

masses on the periodic table on pages 12-13. For example, if you want to compare the molar mass of lead (**Pb**) and iron (**Fe**), you can easily find that lead has more mass per mole – 207.2 g versus 55.8 g.

It's important to realize that mass is not the same thing as size. The two do not go hand-in-hand. In fact, an atom can be larger and actually have less mass; but sometimes a larger atom does have more mass. Therefore, if you want to compare the size of two types of atoms, it is <u>incorrect</u> to simply look at the values of their molar masses.

Size refers to how large something is, which is different from mass. For example, consider a hardcover unabridged dictionary and a standard-size sleeping pillow. The pillow is larger than the dictionary, yet the dictionary has more mass. The dictionary is more dense than the pillow, as more mass occupies less space. A larger atom may or may not be more massive than a smaller atom. So <u>don't</u> look at mass in order to compare the size of two different types of atoms.

Instead, if you want to compare the size of two types of atoms, use the following trends in order to make the comparison. Going down a group, atoms become larger. This is because the additional electrons occupy more distant shells. However, going across a period to the right, atoms become smaller. Although atoms gain electrons going to the right, the atoms become smaller, not larger. Many students find this to seem counterintuitive. The reason that atoms get smaller going to the right is that the nucleus contains more positive charge, and so each electron feels greater attraction to the nucleus, making the size of the atom a little smaller. In contrast to going down the periodic table, going across a period to the right the electrons are added to the same shell, not to a more distant one. This is why the trend for size is different going down than it is going across to the right.

Trends in atomic size

smallest atom

Answer questions based on these trends.
Don't worry about exceptions to the rule.

largest atom

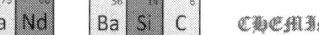 

Which is smaller – carbon (C) or tin (Sn)?

Carbon (C). C is above Sn.

Which is smaller – magnesium (Mg) or sulfur (S)?

Sulfur (S). S is to the right of Mg.

Which is smaller – iron (Fe) or neon (Ne)?

Neon (Ne). Ne is both to the right of and above Fe.

Which is larger – vanadium (V) or nickel (Ni)?

Vanadium (V). The question says larger (not smaller).

Which is smaller – radium (Ra) or Einsteinium (Es)?

Einsteinium (Es). Ra and Es are in the same period.

Ch. 1 Self-Check Exercises

Note: You can check your answers at the back of the book.

1. Indicate if each change is chemical or physical: tea cools in fridge, milk spoils, water boils, hydrogen and oxygen combine to form water, carboard is cut in half.

 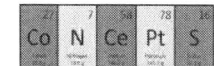
2. Indicate if each substance is an element, compound, solution, or heterogeneous mixture: salt (sodium chloride), milk, lead, paper, tungsten, **MgO**.

3. Give the period and group of each element: helium (**He**), barium (**Ba**), iodine (**I**), aluminum (**Al**).

4. Indicate if each element is a solid, liquid, or gas at STP: hydrogen (**H**), osmium (**Os**), iodine (**I**), mercury (**Hg**).

5. Indicate if each element is a metal or nonmetal: calcium (**Ca**), carbon (**C**), selenium (**Se**), hydrogen (**H**).

6. For each pair of elements, indicate which is more active chemically: calcium (**Ca**)/gallium (**Ga**), nitrogen (**N**)/arsenic (**As**), carbon (**C**)/oxygen (**O**), rubidium (**Rb**)/francium (**Fr**).

7. For each pair of elements, indicate which atom is smaller: titanium (**Ti**)/hafnium (**Hf**), tungsten (**W**)/polonium (**Po**).

8. Order the following atoms from lightest to heaviest: cesium (**Cs**), nitrogen (**N**), iridium (**Ir**), potassium (**K**).

9. Indicate the family that each of the following elements belongs to: bromine (**Br**), sodium (**Na**), xenon (**Xe**), gold (**Au**).

 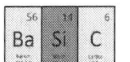 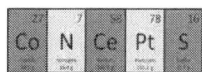
2 S Chemical Bonds

2.1 Electron Structure

85 At Astatine 210.0 g oms are the building blocks of matter because all material objects are composed of different kinds of atoms. Atoms themselves consist of protons, neutrons, and electrons. The protons and neutrons reside in the nucleus of the atom, while electrons orbit in shells and subshells surrounding the nucleus. A sample beryllium (**Be**) atom is illustrated below; beryllium atoms have 4 protons.

Protons are positively charged, neutrons are electrically neutral (they have zero electric charge), and electrons are negatively charged. The electrons are attracted to the protons because opposite charges attract.

Protons and neutrons have approximately the same mass; neutrons are slightly heavier than protons. Electrons are much lighter than both protons and neutrons. The mass of the nucleus approximately equals the mass of the atom.

The electrons of an atom are arranged in shells and subshells surrounding the nucleus. Electron structure is the chief factor in determining

Pa Ge

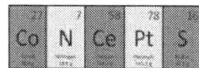

the chemical behavior of an element. Specifically, it is the number of valence electrons that is most important. Valence electrons are the electrons in the outer shell.

Let us first discuss how to determine the number of electrons, and then we will turn our attention to shell structure and valence electrons. A neutral atom has the same number of electrons as protons (otherwise it would be charged, not neutral). Therefore, the total number of electrons (not the number of valence electrons) in a neutral atom equals the atomic number (since atomic number equals the number of protons). Recall that the atomic number is the number in the upper right corner of the element's cell in the periodic table. For example, uranium (**U**) has an atomic number of 92 (see if you can find it on the periodic table).

Determining the total number of electrons in a neutral atom is easy – just find the element on the periodic table and read off the atomic number. As examples, a neutral oxygen (**O**) atom has 8 electrons and a neutral aluminum (**Al**) atom has 13 electrons (note that these are not valence electrons).

The innermost shell of an atom can hold up to 2 electrons, the second and third shells can hold up to 8 electrons each, the fourth and fifth shells can hold up to 18 electrons each, and the sixth and seventh shells can hold up to 32 electrons each.

These 7 shells correspond to the 7 periods (rows) of the periodic table. The first period has 2 elements – hydrogen (**H**) and helium (**He**). (Note: This paragraph – and others – will be much more instructive if you study the periodic table as you read it.) The second and third periods have 8 elements each – lithium (**Li**) thru neon (**Ne**) and sodium (**Na**) thru argon (**Ar**). The fourth and fifth periods have 18 elements each – potassium (**K**) thru krypton

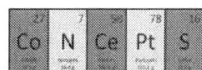

(**Kr**) and rubidium (**Rb**) thru xenon (**Xe**). The sixth and seventh periods have 32 elements each – cesium (**Cs**) thru radon (**Rn**) and francium (**Fr**) thru element 118 (which is yet to be discovered in the laboratory). Remember that the lanthanides – cerium (**Ce**) thru lutetium (**Lu**) – and actinides – thorium (**Th**) thru lawrencium (**Lr**) – are parts of periods 6 and 7, respectively.

Neutral hydrogen (**H**) has just 1 electron, and the first shell of an atom can hold up to 2 electrons. Neutral hydrogen has 1 valence electron, and its outer shell is half full. Neutral helium (**He**) has 2 electrons, so its outer shell (like all noble gases) is completely filled. Neutral lithium (**Li**) has 3 electrons. The first 2 electrons fill the inner shell, and the third electron goes into the second shell. Therefore, neutral lithium has 1 valence electron. Neutral beryllium (**Be**) has 2 valence electrons, boron (**B**) has 3 valence electrons, and if you continue the pattern you will see that fluorine (**F**) has 7 valence electrons. Neon (**Ne**) has a filled outer shell, since its outer shell can hold up to 8 electrons: The first 2 fill the inner shell, and the next 8 fill the second shell.

Neutral sodium (**Na**) has 11 electrons all together. The first 2 fill its innermost shell, the next 8 fill the second shell, and the last 1 is its 1 valence electron. Neutral magnesium (**Mg**) has 2 valence electrons.

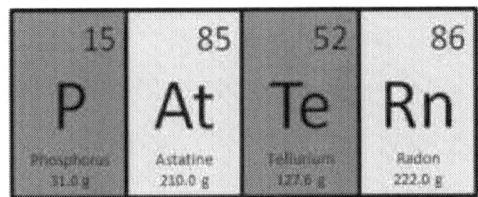

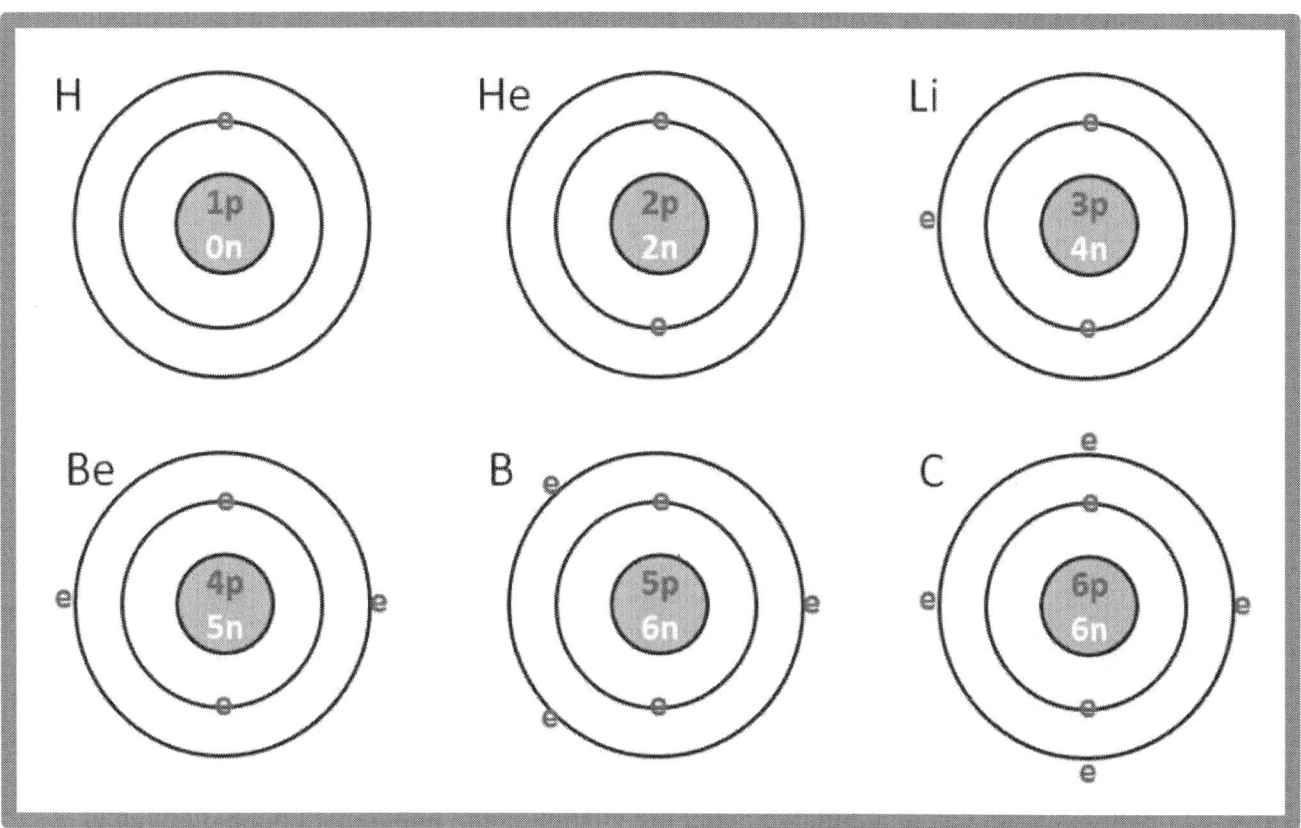

Remember, the first 2 electrons fill the innermost shell, the next 8 go into the second shell, the next 8 go into the third shell, the next 18 go into the fourth shell, the next 18 go into the fifth shell, the next 32 go into the sixth shell, and the next 32 go into the seventh shell. The pattern is 2, 8, 8, 18, 18, 32, 32. You don't need to memorize this: Just look at the periodic table and count the number of elements in each period (row). Try it!

To complicate matters, there are also subshells within shells, and these have some subtle effects among the transition metals (between groups 2 and 3 – see page 16) and the metals in groups 3 thru 6. However, the main ideas of how most of the metal and nonmetal atoms behave can be understood by examining the outermost shell. Therefore, we will focus our attention on counting valence electrons in the outer shell, and not worry about subshells until Chapter 6.

Shell Structure
Strontium (Sr)

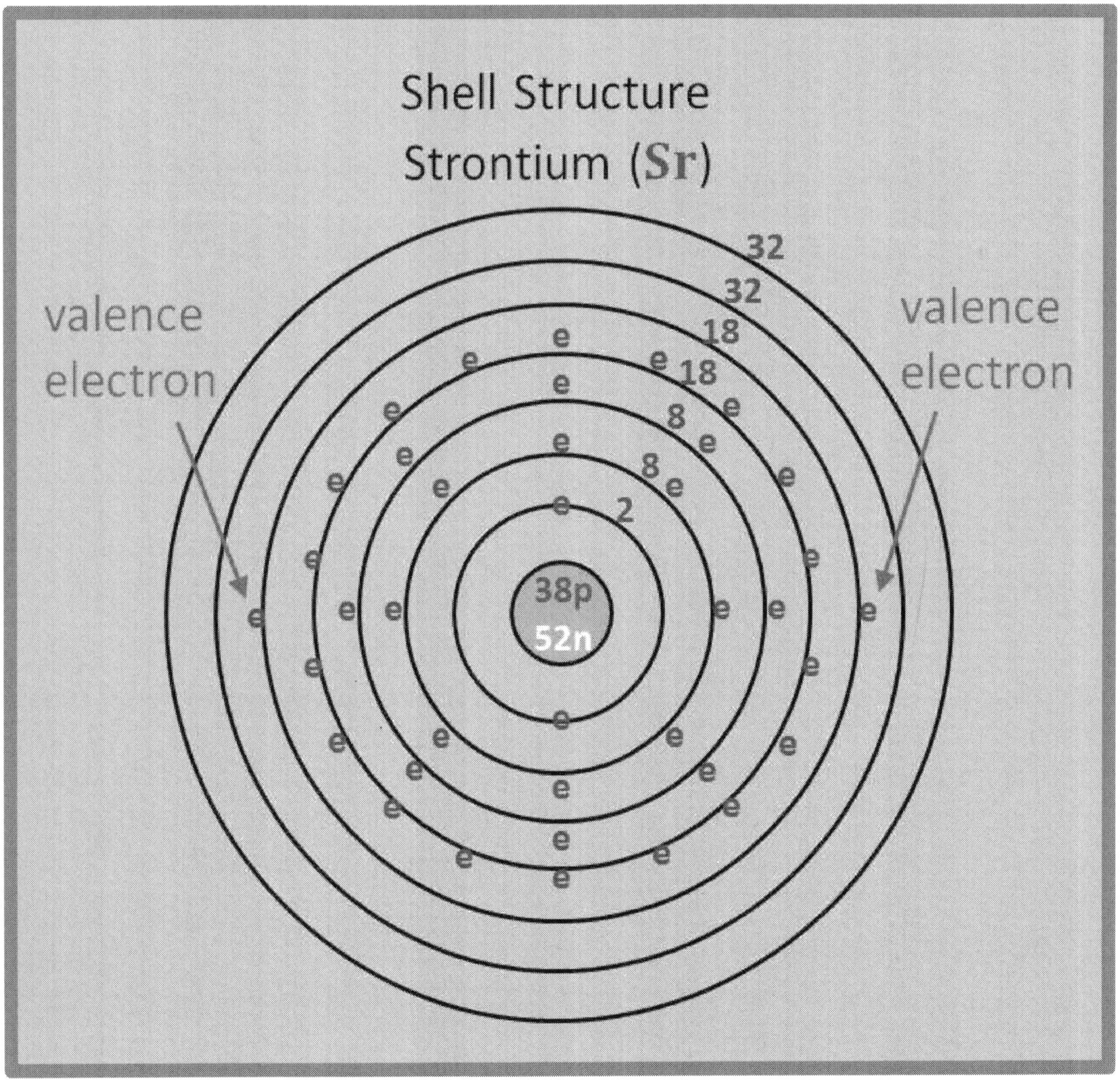

valence
electron

valence
electron

32
32
18
18
8
8
2

38p
52n

e (multiple, arranged in shells)

Since the number of electrons in each shell corresponds to the number of elements in a period, it's easy to count valence electrons: Just count from the left. For example, in period 3, neutral sodium (**Na**) has 1 valence electron, magnesium (**Mg**) has 2, aluminum (**Al**) has 3, and if you continue the pattern you will see that chlorine (**Cl**) has 7. Remember, valence electrons are the electrons in the outermost shell.

How many electrons are there in a neutral zinc (Zn) atom?

- There are **30** electrons all together.
- The question does not say valence electrons.

How many valence electrons are there in a neutral magnesium (Mg) atom?

- There are **2** electrons in its outer shell.
- Count from the left of its row: **Na** 1, **Mg** 2.

How many valence electrons are there in a neutral oxygen (O) atom?

- There are **6** electrons in its outer shell.
- Count from the left of its row: **Li** 1, **Be** 2, **B** 3, **C** 4, **N** 5, **O** 6.

Why are valence electrons so important? Because atoms tend to gain, lose, or share their valence electrons in such a way as to have a filled outer shell. For example, the distinction between metal and nonmetal atoms is that metal atoms tend to lose valence electrons while nonmetal atoms tend to gain them. We will explore this in much more detail in the remaining sections of this chapter.

2.2 Metal and Nonmetal Ions

hen a neutral atom gains or loses electrons, it becomes electrically charged. We call a charged atom an ion. A neutral atom that gains one or more electrons becomes a negatively charged ion (since electrons are negative), while a neutral atom that loses one or more electrons becomes a positively charged ion.

Ions are atoms with more or fewer electrons than protons.

- A neutral atom has as many electrons as protons.

- An ion either has more electrons than protons or fewer electrons than protons.

- A neutral atom can become an ion by gaining or losing electrons.

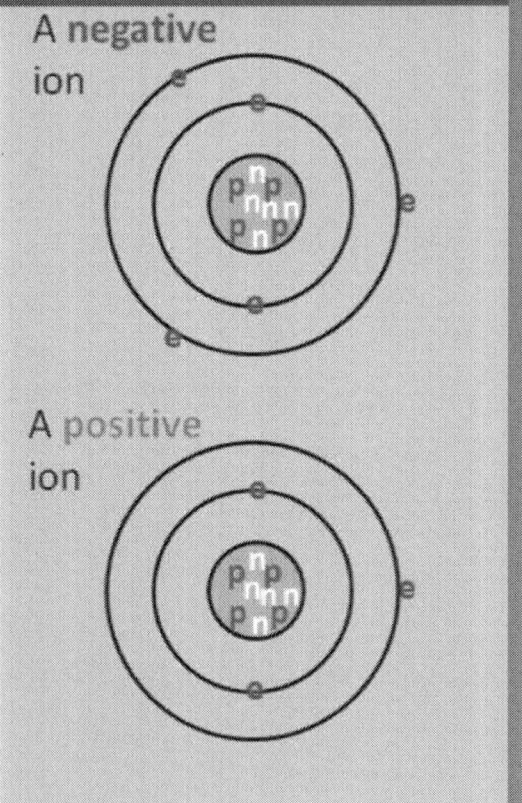

A negative ion

A positive ion

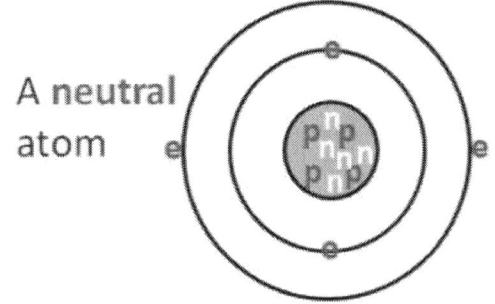

A neutral atom

Ions are formed by gaining or losing electrons – not protons. The protons are much heavier and reside in the nucleus, whereas the electrons

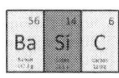 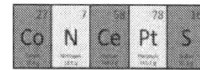
are much lighter and surround the atom. Lighter particles are easier to accelerate, and particles further from the center of the atom require less energy to release them. Finally, the number of protons determines the element. If the number of protons were to change, it would change the identity of the element. For example, if a carbon (**C**) atom gains or loses electrons, it still has 6 protons and therefore is still a carbon atom (only now it is a charged atom, so we call it a carbon ion). However, if a carbon atom were to gain or lose a proton, it wouldn't be a carbon atom anymore. Boron (**B**) has 5 protons, nitrogen (**N**) has 7 protons, and so on. The number of protons (not electrons or neutrons) determines the identity of the element.

The electrons that are easiest to remove are the valence electrons because they are in the outer shell of the atom. Since the valence electrons are the electrons involved in forming ions and are also the electrons involved when multiple atoms share electrons, they are very important for understanding chemical behavior.

When an atom loses electrons, it loses valence electrons. When an atom gains an electron, that electron becomes a valence electron. Boron (**B**) has 5 protons. Therefore, neutral boron has a total of 5 electrons: 2 reside in the inner shell, while the other 3 are valence electrons. The **B**$^+$ ion is a boron atom that has lost 1 valence electron – so it has 2 valence electrons. The **B**$^-$ ion is a boron atom that has gained 1 valence electron – so it has 4 valence electrons.

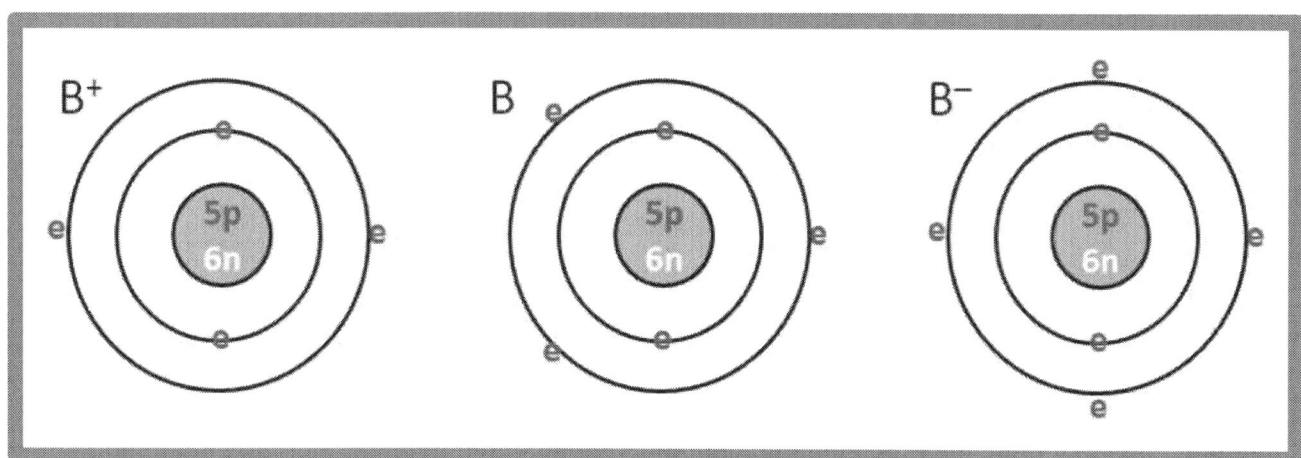

A superscript with a + or − sign is used to denote an ion. When an element appears without a + or − superscript, it designates a neutral atom. For example, **H** (hydrogen), **Al** (aluminum), **Cl** (chlorine), and **O** (oxygen) are neutral atoms.

If a neutral Hg atom gains an electron, will it become a positive or negative ion?

It will become a negative ion because it will have more electrons than protons.

If a neutral Hg atom loses an electron, will it become a positive or negative ion?

It will become a positive ion because it will have fewer electrons than protons.

A + or − superscript designates a positive or negative ion. A number with the + or − indicates how many electrons were lost or gained, respectively. Here are a few examples:

- H^+ is a positive hydrogen (**H**) ion: It has 1 proton, but no electrons. H^+ is an **H** atom that lost its 1 electron.

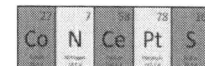

- Al^{3+} is a positive aluminum (**Al**) ion. It's an **Al** atom that has lost all 3 of its valence electrons.
- Cl^- is a negative chlorine (**Cl**) ion. It's a **Cl** atom that has gained 1 electron.
- O^{2-} is a negative ion. It has 2 more electrons than protons.

How many protons and electrons are there in a neutral beryllium (Be) atom?

- The number of protons equals the atomic number – **four**.
- It's neutral, so there are just as many electrons as protons – **four**.

How many protons and electrons are there in a Be^- ion?

- It still has **four** protons.
- It has one extra electron – so **five**.

How many protons and electrons are there in a Be^{2+} ion?

- It still has **four** protons.
- It has two less electrons – so **two**.

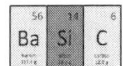 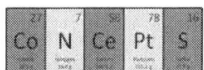

Atoms have a natural tendency to have a filled outer shell. One way for an atom to have a filled outer shell is to gain or lose electrons, forming an ion. For example, a neutral potassium (**K**) atom has 1 valence electron and a neutral chlorine (**Cl**) atom has 7 valence electrons. Chlorine needs 1 more electron to fill its outer shell, since its outermost shell can hold up to 8 electrons. When potassium and chlorine atoms get together, each potassium atom likes to donate one electron to a chlorine atom, forming the ions **K$^+$** and **Cl$^-$**. These oppositely charged ions then attract, forming an ionic bond. We will discuss ionic bonds in the next section; for now, let us focus on which ions metal and nonmetal elements tend to form.

How many valence electrons are there in a neutral carbon (C) atom?

- There are **four** electrons in its outer shell.
- Count from the left of its row: Li 1, Be 2, B 3, C 4.

How many valence electrons are there in a C^{2-} ion?

- Compared to neutral carbon, this ion has two extra electrons – so **six**.

How many valence electrons are there in a C$^+$ ion?

- Compared to neutral carbon, this ion has one less electron – so **three**.

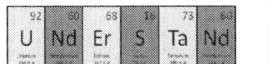 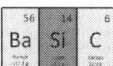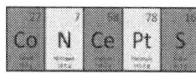

Metals lie on the left side of the zigzag staircase on the periodic table, whereas nonmetals lie on the right side of the zigzag staircase. The underlying distinction between metals and nonmetals is that a metal atom prefers to lose electrons to have a filled outer shell, while a nonmetal atom prefers to gain electrons to have a filled outer shell. These tendencies match the structure of their valence electrons. Most metals just have a few valence electrons, so it's much easier to lose a small number of electrons – just one for the alkali metals, two for the alkaline earth metals, and a few for other metals – than it is to gain several electrons.

Beryllium (**Be**), for example, only needs to lose 2 electrons in order to have a filled outer shell, whereas it would need to gain 6 (since its outer shell can hold up to 8). It's much more likely to find a partner to accept 2 electrons than to give up 6 (and less energy is required to do so). Fluorine (**F**), on the other hand, only needs to gain 1 electron in order to have a filled outer shell (since it has 7 valence electrons and its outer shell can hold up to 8). Thus, fluorine is much more apt to gain 1 electron than to lose all 7.

Noble gases don't like to gain or lose electrons. Why not? Because their outer shells are already full. Noble gases, like helium (**He**) and neon (**Ne**) are generally found as neutral atoms, not as ions. Metals and the other nonmetals are not always found as ions; they are sometimes found as neutral elements. However, they often form ions and often bond with other elements ionically or covalently (see the next two sections) in order to have

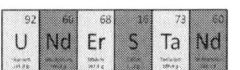 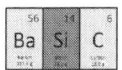
filled outer shells. Elements have a natural tendency to behave this way, but they don't always have the opportunity – the electrons have to go or come from somewhere, so they need a willing partner to exchange or share electrons with.

Metals readily lose one or more valence electrons, becoming positive ions like Li^+ or Ca^{2+}. Nonmetals readily gain one or more electrons, becoming negative ions like O^{2-} or F^- – except for noble gases, which prefer to remain as neutral atoms.

The "natural" ion of an atom is the ion that the atom tends to form in order to have a filled outer shell. In forming the natural ion, a metal atom loses electrons, while a nonmetal atom gains electrons, until the outer shell is filled.

For a metal, count from the element to the left on the periodic table in order to determine its natural ion, which will be positive. Include the metal when you count. The reason for this is that the metal's position from the left on the periodic table equals its number of valence electrons. A metal atom must lose all of its valence electrons in order for its outer shell to be full.

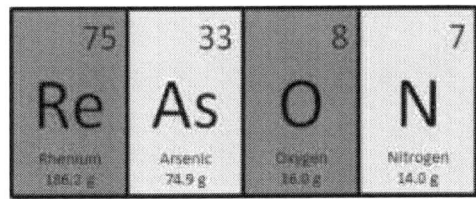

For example, to find the natural ion for barium (**Ba**), which is a metal, count to the left starting with **Ba**: 1 for **Ba**, 2 for **Cs** (cesium). Barium has 2 valence electrons. Barium tends to form the ion Ba^{2+}. (Instead of counting to the left, you can just use the number of valence electrons, which is the same thing.)

Many of the transition metals (between groups 2 and 3 – see page 16) and the metals in groups 3 thru 6 don't follow this rule because they have a more complicated shell and subshell structure. We will find a table of exceptions for the transition metals in the following section. For all other metals, we will apply this rule.

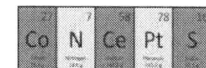
For a nonmetal, count from the element to the right in order to determine its natural ion, which will be a negative ion. Don't include the nonmetal when you count. We are counting to the right because a nonmetal atom tends to gain electrons. The rightmost element (a noble gas) has a filled outer shell.

For example, to find the natural ion for sulfur (S), which is a nonmetal, count to the right – but this time don't start with the element itself, S. Instead, we count 1 for Cl (chlorine), 2 for Ar (argon). Sulfur has 6 valence electrons; it needs 2 more electrons in order to fill its outer shell. Sulfur tends to form the ion S^{2-}. (Instead of counting to the right, you can just subtract the number of valence electrons from the total number of elements in the period, which is the same thing. For example, there are 8 elements in sulfur's period and sulfur has 6 valence electrons: $8 - 6 = 2$.)

Following are a few more examples of how to determine the natural ion of an atom. As with most of the examples in this book, you will find them to be more instructive if you study the periodic table as you study them.

- Radium (Ra) is a metal. Count 1 for Ra, 2 for Fr (or simply count that Ra has 2 valence elctrons). You get Ra^{2+}.
- Francium (Fr) is a metal. Count 1 for Fr (or just count Fr's single valence electron). You get Fr^{+}.
- Astatine (At) is a nonmetal. Don't count At, count 1 for Xe (or subtract 31 from 32 since At has 31 electrons in its sixth shell, which can hold up to 32). You get At^{-}.
- Oxygen (O) is a nonmetal. Don't count O, count 1 for F, count 2 for Ne (or subtract O's 6 valence electrons from 8, since its outer shell can hold 8). You get O^{2-}.

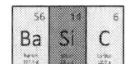 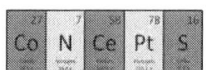

What is the natural ion of potassium (K)?

K^+. Count 1 for K.

What is the natural ion of calcium (Ca)?

Ca^{2+}. Count 1 for Ca and 2 for K.

What is the natural ion of aluminum (Al)?

Al^{3+}. Count 1 for Al, 2 for Mg, and 3 for Na.

What is the natural ion of bromine (Br)?

Br^-. Don't count Br, count 1 for Kr.

What is the natural ion of phosphorus (P)?

P^{3-}. Don't count P, count 1 for S, 2 for Cl, and 3 for Ar.

2.3 Ionic Bonds

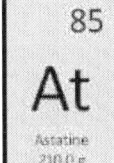 oms bond with one another in two different ways – through ionic bonds or through covalent bonds. When a metal transfers one or more electrons to a nonmetal, they form an ionic bond. When two nonmetals share one or more electrons, they form a covalent bond. We will discuss ionic bonds in this section and covalent bonds in the next section.

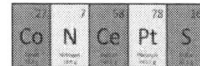
Electrons are exchanged in ionic bonds, causing the atoms to form ions. Hence the name ionic.

Types of bonds

- When two nonmetals **share** one or more electrons, they form a **covalent bond**.
- When a metal **transfers** one or more electrons to a nonmetal, they form an **ionic** bond.
- When electrons are exchanged, the atoms form ions; hence the name ionic.

covalent (2 nonmetals)

$$H \underset{e^-}{\overset{e^-}{—}} H$$

$$H_2$$

ionic (metal-nonmetal)

$$e^- \; e^- \; e^-$$
$$e^- Na^+ e^- \quad e^- Cl^- e^-$$
$$e^- \; e^- \; e^-$$

NaCl

53	8	28	6
I	O	Ni	C
Iodine	Oxygen	Nickel	Carbon
126.9 g	16.0 g	58.7 g	12.0 g

Metal atoms bond with nonmetal atoms ionically. The reason for this is that a metal atom prefers to lose its valence electrons in order to have a filled outer shell, while a nonmetal atom prefers to gain electrons in order to fill its outer shell. It's a match made in Heaven!

When metal atoms interact with nonmetal atoms, the metal atoms donate their valence electrons to the nonmetal atoms. The metal atoms become positively charged ions and the nonmetal atoms become negatively charged ions. These ions attract one another because they are oppositely charged, forming an ionic bond.

When a metal transfers one or more electrons to a nonmetal, they form an ionic bond. Sodium (**Na**) has one extra electron, while chlorine (**Cl**) needs one electron. Sodium donates an electron to chlorine, forming the ionic compound sodium chloride (**NaCl**) – a bond between Na^+ and Cl^- ions. Observe that chlorine's suffix changes from –*ine* to –*ide* when naming the compound. Chlorine refers to the individual **Cl** atom, whereas sodium chloride is a new substance, **NaCl**. It's important to realize that sodium chloride is a considerably different substance than just sodium and chlorine mixed together – it is a chemical change, not simply a physical mixture. We will discuss the rules for how to properly name compounds in the next chapter. For now, it will be useful to note that the end of the nonmetal generally changes to –*ide* when naming an ionic compound.

Ionic bonds

- When a metal **transfers** one or more electrons to a nonmetal, they form an **ionic** bond.

- Sodium (Na) has one extra electron, while chlorine (Cl) needs one electron. Sodium donates an electron to chlorine, forming the ionic compound sodium chloride (NaCl) – a bond between Na^+ and Cl^- ions.

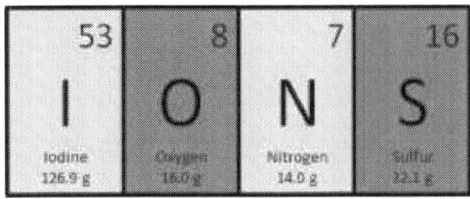

e^- Be^{2+} e^-

e^- e^- e^- e^- e^- e^-
e^- F^- e^- e^- F^- e^-
e^- e^- e^- e^- e^- e^-

BeF_2

e^- e^- e^- e^- e^- e^-
e^- Na^+ e^- e^- Cl^- e^-
e^- e^- e^- e^- e^- e^-

$NaCl$

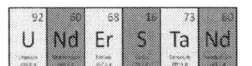 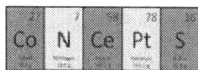

In forming sodium chloride ($NaCl$), every sodium (Na) atom loses one electron to form Na^+ and every chlorine (Cl) atom gains one electron to form Cl^-. There are just as many Na^+ ions as there are Cl^- ions.

Compare this to calcium fluoride (CaF_2). Calcium (Ca) has two valence electrons to lose, but fluorine (F) only needs to gain one electron to fill its outer shell. In this case, each Ca atom bonds with two F atoms, forming Ca^{2+}, F^-, and F^-. The chemical formula CaF_2 has a subscript 2 because there are twice as many F^- ions as there are Ca^{2+} ions.

Following are the rules for how to determine the chemical formula for an ionic bond:

1. First, write the natural ions (as described in the previous section). For example, magnesium's natural ion is Mg^{2+} and chlorine's natural ion is Cl^-.

2. Next, swap the superscripts to make the subscripts. If there are any 1's, don't write them. For example, Mg^{2+} and Cl^- combine to form $MgCl_2$. The superscript $2+$ from Mg^{2+} became the subscript 2 for Cl_2, while the superscript $-$ from Cl^- became the subscript 1 for Mg (but the 1's are implied – i.e. we don't write 1's in the formulas).

3. Finally, check if your answer can be reduced. For example, Ca_2O_2 becomes CaO. When you cross the superscripts in Ca^{2+} and O^{2-} to make the subscripts in Ca_2O_2, this means that there are 2 Ca^{2+} ions for every 2 O^{2-} ions. This means that there is one Ca^{2+} ion for every O^{2-} ion. CaO is simpler than Ca_2O_2.

Following are a few examples for how to determine the chemical formula of an ionic compound:

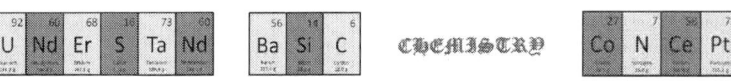

- Lithium (**Li**) and fluorine (**F**). Their natural ions are Li^+ and F^-. Cross the superscripts to make **LiF**.

- Calcium (**Ca**) and chlorine (**Cl**). Their natural ions are Ca^{2+} and Cl^-. Cross the superscripts to make $CaCl_2$.

- Sodium (**Na**) and oxygen (**O**). Their natural ions are Na^+ and O^{2-}. Cross the superscripts to make Na_2O.

- Magnesium (**Mg**) and sulfur (**S**). Their natural ions are Mg^{2+} and S^{2-}. Cross the superscripts to make Mg_2S_2. This reduces to **MgS**.

Not all ionic bonds form from a single metal ion and a single nonmetal ion. Sometimes the ion of a polyatomic atom group is involved in the bond. For example, when potassium (**K**) interacts with hydrogen (**H**) and oxygen (**O**), the potassium atom can form an ionic bond with a hydroxide (OH^-) ion. The hydroxide ion is a covalent bond (described in the following section) between two nonmetals, while the bond between K^+ and OH^- is ionic.

Each **O** atom needs 2 electrons to fill its outer shell and each **H** (which is a nonmetal) atom needs 1 electron to fill its outer shell (since **H** just has one shell, and the first shell can hold 2 electrons). If hydrogen and oxygen interact just among themselves, they like to form water (H_2O), which is a covalent bond that we will study in the following section. However, when hydrogen and oxygen interact with a metal, they may form the hydroxide (OH^-) ion and bond ionically with the metal.

Water (H_2O) molecules are neutral. Each hydrogen atom shares 1 electron with oxygen. The shared electron makes hydrogen feel like it has 2 electrons (its own plus the shared electron), while oxygen feels like it has 8 valence electrons (its own 6 valence electrons plus 2 shared electrons). All three atoms (2 **H** and 1 **O**) feel like their outer shells are full.

Write the chemical formula for potassium chloride.

1. K^+, Cl^-. 2. KCl.

Write the chemical formula for cesium sulfide.

1. Cs^+, S^{2-}. 2. Cs_2S.

Write the chemical formula for barium fluoride.

1. Ba^{2+}, F^-. 2. BaF_2.

Write the chemical formula for radium oxide.

1. Ra^{2+}, O^{2-}. 2. Ra_2O_2. 3. RaO.

Write the chemical formula for aluminum sulfide.

1. Al^{3+}, S^{2-}. 2. Al_2S_3.

In contrast to water, the hydroxide (OH^-) atom group is a negative ion. There is only one **H** atom in hydroxide. When a single **H** and **O** interact, the **H** atom shares one electron with **O**, which makes **H** happy, but **O** needs 1 more electron to feel like its outer shell is full. The **H-O** pair tends to form hydroxide by accepting an electron from a metal. In this way, the polyatomic ion hydroxide (OH^-) bonds ionically with metal atoms.

When potassium interacts with hydrogen and oxygen, the potassium atom donates one electron to the hydrogen-oxygen pair, forming the K^+ ion and the hydroxide (OH^-) ion. These oppositely charged ions then attract to one another, forming the ionic bond potassium hydroxide (**KOH**).

There are many other polyatomic ions besides the hydroxide ion. Some common polyatomic ions are tabulated on the following page.

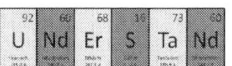 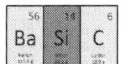

The formulas for the ionic bonds involving polyatomic ions work the same way as the ionic bonds between metal and nonmetal atoms, except that you can just find the ion of the polyatomic atom in the table below. For example, when aluminum (Al) combines with nitrate (NO_3^-), we first write the natural ion of aluminum, which is Al^{3+}, then we cross the superscripts to form the subscripts as usual, forming the ionic bond aluminum nitrate, for which the formula is $Al(NO_3)_3$. The parentheses with the additional subscript 3 means that there are 3 NO_3^- ions for each Al^{3+} ion.

Common polyatomic ions

ammonium	NH_4^+	KNO_3
carbonate	CO_3^{2-}	potassium nitrate
cyanide	CN^-	$Ca(OH)_2$
hydroxide	OH^-	calcium hydroxide
nitrate	NO_3^-	$Al_2(CO_3)_3$
nitrite	NO_2^-	aluminum carbonate
phosphate	PO_4^{3-}	$(NH_4)_2SO_3$
sulfate	SO_4^{2-}	ammonium sulfite
sulfite	SO_3^{2-}	

20	37	8	11	52
Ca	Rb	O	Na	Te
Calcium 40.1 g	Rubidium 85.5 g	Oxygen 16.0 g	Sodium 23.0 g	Tellurium 127.6 g

As another example, consider the ionic bond that forms when the ammonium ion (NH_4^+) – not to be confused with neutral ammonia atoms (NH_3) – interacts with sulfur (S). The natural ion of sulfur is S^{2-}. When we

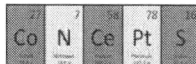

cross the superscripts to make the subscripts, we get $(NH_4)_2S$ as the formula for ammonium sulfide. See if you can understand the four chemical formulas near the table of polyatomic ions on the previous page.

The transition metals (between groups 2 and 3 of the periodic table – see page 16) and the metals in groups 3 thru 6 (but not aluminum) have a complicated subshell structure. Many of these elements form multiple ions because they can lose electrons from two or more different subshells, filling their outer subshells in different ways. They often lose just the electrons of a subshell, and not the entire shell. These metals do not follow the usual rule for determining the natural ion. Instead, you can find a table of many of their common ions below.

Common transition metal ions

copper	Cu^+, Cu^{2+}
Iron	Fe^{2+}, Fe^{3+}
lead*	Pb^{2+}, Pb^{4+}
mercury	Hg^+, Hg^{2+}
nickel	Ni^{2+}
silver	Ag^+
tin*	Sn^{2+}, Sn^{4+}
zinc	Zn^{2+}

CuO
copper (II) oxide

Cu_2O
copper (I) oxide

SnI_4
tin (IV) iodide

HgS
mercury (II) sulfide

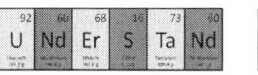 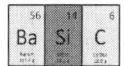 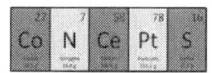

For example, consider iron (**Fe**), which is atomic number 26 on the periodic table. According to the usual rule, you would expect the number of valence electrons to be 8, since it is the eighth element from the left. However, iron is a transition metal with a complicated subshell structure. Instead, iron commonly forms the ions Fe^{2+} and Fe^{3+}. When you write the formula for an ionic bond, if the metal is a transition metal or to the right of the transition metals, you need to consult the table of common transition metal ions on the previous page in order to determine the formula. We will consider this again in the following chapter, when we discuss how to name ionic bonds when a transition metal is present.

The formulas for the ionic bonds involving transition metals (between groups 2 and 3 – see page 16) or metals in groups 3 thru 6 (but note that aluminum follows the usual rules) work the same way as the ionic bonds between group 1 or 2 metal atoms and nonmetal atoms, except for using the previous table to determine the common ion for the metal. For example, when zinc (**Zn**) and fluorine (**F**) interact, zinc typically forms the Zn^{2+} ion listed in the previous table, while the natural ion of fluorine is F^-. We cross the superscripts of the ions as usual to form the subscripts of zinc (II) fluoride: ZnF_2. The Roman numeral II in parentheses refers to the charge of the zinc ion. We include a Roman numeral when naming ionic compounds involving a metal to the right of group 2 (except for aluminum) in order to avoid ambiguity, as we will explain in the next chapter.

As another example, consider the ionic bond formed between the lead ion Pb^{4+} and the natural oxygen ion O^{2-}. When we cross the superscripts to make the subscripts, we get Pb_2O_4, which reduces to PbO_2.

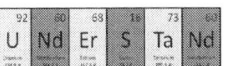 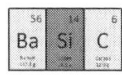 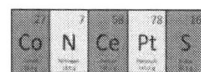
2.4 Covalent Bonds

 valent bonds involve two (or more) nonmetals sharing electrons. It is important to realize that the individual atoms are neutral – they are not ions, so they are not charged – and that the electrons are shared rather than transferred in covalent bonds (unlike ionic bonds).

Although the covalent bond is quite different in character from the ionic bond – shared electrons versus transferred electrons – both types of bonds are based on the same underlying concept: Atoms tend to behave in such a way as to effectively fill their outer shells. In the case of an ionic bond, each atom fills its shell by losing or gaining valence electrons. In the case of a covalent bond, the atoms share electrons so that each atom effectively fills its outer shell.

A single water (H_2O) molecule consists of a covalent bond between one oxygen (O) atom and two hydrogen (H) atoms. The oxygen atom has 6 valence electrons and needs 2 more electrons to fill its outer shell. Each hydrogen atom has 1 valence electron and needs 1 more electron to fill its outer shell (since there are only 2 elements in the first period). Oxygen shares a pair of electrons with each hydrogen atom. Hydrogen shares its original electron and one from oxygen. In this way, each hydrogen atom feels the presence of two electrons in its outer shell. Oxygen had 6 electrons originally, and feels the presence of the 2 electrons that the hydrogen atoms are sharing with oxygen. All three atoms in one water molecule thus feel the effect of having a filled outer shell by sharing their electrons in this way. See the figure on the following page.

Covalent bonds

- When two nonmetals **share** one or more electrons, they form a **covalent bond**.

- Two hydrogen (H) atoms each need 1 electron to fill their outer shell, so they each share their electrons to have a sense of a filled outer shell.

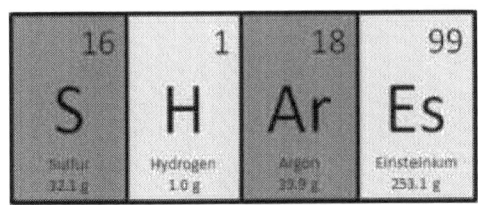

Methane (CH_4) has one carbon (**C**) atom and four hydrogen (**H**) atoms. The carbon atom has 4 valence electrons and needs 4 more electrons in order to fill its outer shell. Since each hydrogen has 1 valence electron and needs 1 more to fill its outer shell, each hydrogen atom shares a pair of electrons (its own plus one from carbon) with the carbon atom. Carbon feels the presence of 8 electrons (4 of its own plus 1 each from the 4 hydrogen atoms). See the methane molecule in the previous figure.

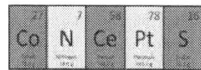

Sometimes the atoms of a single element bond together in pairs covalently. This is the case, for example, in diatomic gases. For example, hydrogen (**H**) atoms form a diatomic gas, **H$_2$**, at standard temperature and pressure (STP). The subscript 2 indicates that the hydrogen atoms are bonded together in pairs – in contrast to a monatomic gas that is a collection of individual atoms. Two hydrogen atoms share their combined 2 electrons such that each hydrogen atom feels the effect of having a filled shell.

Helium (**He**), a noble gas, already has a filled outer shell. Helium gas is monatomic, as the atoms remain separate – they do not bond together in pairs like the molecules of a diatomic gas.

Monatomic and diatomic molecules

- Helium gas consists of individual helium atoms. Helium gas is **monatomic** (He).

- Hydrogen gas consists of pairs of hydrogen atoms bonded together covalently. Hydrogen gas is **diatomic** (H$_2$).

- **H, N, O, F, Cl, Br, and I tend to be diatomic. These last six form a 7 on the periodic table (next slide).**

 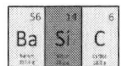

How do nitrogen (N) and hydrogen (H) share electrons in the covalent bond NH₃?

- Hydrogen **needs one** electron to fill its outer shell.
- Nitrogen **needs three** electrons to fill its outer shell.
- NH₃ consists of 1 N atom and 3 H atoms.
- Therefore, each H atom shares its electron with the N atom, and the N atom shares 3 of its electrons with the H's.
- This way, the N atom feels like it has 8 electrons, and each H atom feels like it has 2 electrons.

The elements that tend to form diatomic molecules include hydrogen (**H**), nitrogen (**N**), oxygen (**O**), fluorine (**F**), chlorine (**Cl**), bromine (**Br**), and iodine (**I**). This turns out to be very easy to memorize, as the last six of these elements form the shape of a 7 on the periodic table (as you can see in the following periodic table). The right side of this 7 is in group 7 (the halogens) and the top of the 7 is in period 2 (row 2).

31	16	99
Ga	S	Es
Gallium	Sulfur	Einsteinium
69.7 g	32.1 g	253.1 g

91	32
Pa	Ge

Diatomic molecules

monatomic

diatomic

Yes, you need to remember which are diatomic.

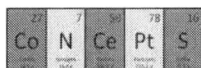

Carbon (**C**), hydrogen (**H**), and silicon (**Si**) are special elements in that they have half as many valence electrons as they need to fill their outer shells: Hydrogen has 1 valence electron and needs 1 more, while carbon and silicon each have 4 valence electrons and need 4 more. For one thing, this allows these elements to easily bond with themselves. For example, diamond and graphite are different arrangements of carbon atoms. For another, these elements tend to be involved in a large number of different covalent bonds. The wide variety of complex molecules that make up living organisms, for example, contain carbon. The chemistry of carbon is so rich – and so central to biology – that there is an entire branch devoted to the chemistry of carbon called organic chemistry. Silicon has many practical uses in electronics and computer chips.

2.5 The Outer Shell

hy is electron structure so important for chemistry? Because much of the chemical behavior of elements stems from the fact that atoms like to have their outer shells filled.

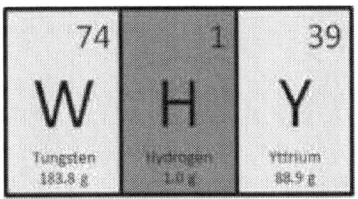

There are two main classes of elements because there are two different ways for atoms to fill their outer shells. Atoms that have nearly full outer shells like to gain electrons in order to fill them – either by accepting them from other atoms or by sharing them. Elements that tend to gain electrons are called nonmetals. Atoms with only a few electrons in their outer shell (or subshell, in the case of the transition metals and metals in groups 3 thru 7

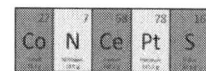

below aluminum) like to lose their electrons in order to have filled outer shells. Elements that tend to lose valence electrons are called metals. It turns out that the majority of elements are metals. See the periodic table on page 21.

Atoms bond together in two ways – ionically or covalently. Both types of bonds are based on the fact that atoms like to have filled outer shells. In ionic bonds, metal atoms donate all of their valence electrons to nonmetal atoms (or there may be polyatomic ions involved instead), and all of the atoms – metals and nonmetals alike – get their outer shells filled in the process. After the exchange, the oppositely charged ions attract, forming an ionic bond.

The chemical formula is dictated by the number of valence electrons that each metal atom has and each nonmetal atom needs to fill its outer shell. For example, sodium chloride (**NaCl**) and magnesium oxide (**MgO**) do not have any subscripts because the donor has exactly what the recipient needs: Sodium (**Na**) has 1 valence electron and chlorine (**Cl**) needs 1, and magnesium (**Mg**) has 2 valence electrons and oxygen (**O**) needs 2. Compare this with sodium oxide (**Na$_2$O**) and calcium chloride (**CaCl$_2$**), which have subscripts because the balance is uneven: Sodium (**Na**) has 1 valence electron, but oxygen (**O**) needs 2, and calcium (**Ca**) has 2, but chlorine (**Cl**) only needs 1. Therefore, two sodium join with one oxygen and two chlorine join with one calcium in these respective ionic bonds.

Nonmetals bond with other nonmetals covalently by sharing electrons so that both atoms have a sense of having a filled outer shell. For example, two hydrogen (**H**) atoms bond with one oxygen (**O**) atom in a single water (**H$_2$O**) molecule by sharing electrons in a way that effectively fills their outer shells. Each hydrogen atom shares a pair of electrons with the oxygen atom. Each hydrogen feels the presence of 2 electrons in its only shell (its own electron plus one from oxygen), and oxygen feels the presence of 8 electrons in its outer shell (its own 6 valence electrons plus the 2 it is sharing with the hydrogen atoms).

All of the elements in a group (column) of the periodic table have similar chemical behavior because they have the same number of valence

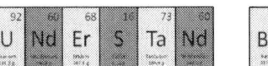 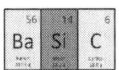
electrons. For example, the noble gases (group 8) react very little, whereas the halogens (group 7) are highly reactive nonmetals and the alkali metals are highly reactive metals (group 1).

Ch. 2 Self-Check Exercises

1. Determine the total number of electrons in each of the following: He, Mg, Cu, U, H^+, Ca^+, Fe^{2+}, Sn^{3+}, Fr^+, H^-, O^{2-}, S^-, Br^-, Te^{2-}.

2. Determine the number of <u>valence</u> electrons in each of the following: He, Na, Mg, Al, Cl, K^+, Mg^{2+}, Al^{2+}, Ba^{2+}, Ra^+, O^-, N^{2-}, P^{3-}, Ar^-, Cl^-.

3. Indicate whether each of the following bonds is mostly covalent or mostly ionic: $MgCl_2$, H_2O, F_2, Na_2O, N_2O_3, Fe_2O_3, CO_2.

4. Indicate whether each of the following elements tends to form monatomic or diatomic molecules: hydrogen (H), sulfur (S), helium (He), bromine (Br), argon (Ar).

5. Indicate the natural ions for each of the following elements: Li, N, Ba, Se, Sr, F, O, Be, Al, P, Cl, Cs, S, Rb.

6. Write the chemical formula for the following bonds: Li-F, Be-Cl, K-S, Mg-N, Ca-S, Al-O, Sr-O, Al-F, Fr-Cl, Sr-N.

7. Write the chemical formula for the covalent bond that hydrogen (H) and sulfur (S) are likely to make. Draw a picture and show how the hydrogen and sulfur atoms share electrons in this bond.

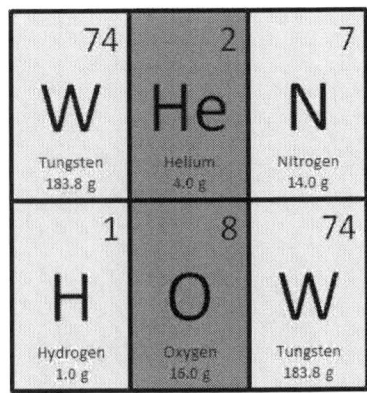

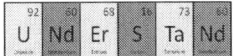 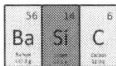
3 ░ Naming Compounds

3.1 Naming Scheme

ot all compounds are named similarly. For example, dinitrogen trioxide (N_2O_3) has prefixes *di-* and *tri-* while calcium chloride ($CaCl_2$) does not, but iron (III) oxide (Fe_2O_3) includes a Roman numeral III. Some compounds, like water (H_2O) and ammonia (NH_3) have special names. Acids, like hydrochloric acid (HCl), are also named differently. Fortunately, there is a set of rules that you can learn in order to speak the language of chemistry fluently. We will discuss the rules for naming compounds in this section, and then address each rule individually in the subsequent sections with additional examples.

First, there is one rule that applies to almost all binary compounds. A binary compound is a compound that consists of two elements, like sodium chloride (NaCl) or carbon dioxide (CO_2). When a compound consists of two elements, the second element almost always ends with the suffix *-ide*. For example, the compound Al_2O_3 is called aluminum oxide.

The last element of a binary compound is always a nonmetal, regardless of whether the bond is ionic or covalent. Instead of just writing the name of the second element as its usual name, we change its suffix to -

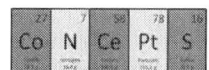

ide, which reinforces the notion that the resulting compound is a much chemically different substance than the original two elements. For example, when sodium (**Na**) and chlorine (**Cl**) react together to form sodium chloride (**NaCl**), the new compound is significantly different from either sodium or chlorine – this chemical reaction involves a chemical change from two substances to a brand new substance. Hence, we call **NaCl** sodium chloride (instead of sodium chlorine).

The table below lists the name change for several nonmetals when they appear at the end of a binary compound.

Binary compound endings

When a compound consists of **two elements**, the **second element ends with -ide.** Some common endings:

hydride	nitride	oxide
fluoride	sulfide	chloride
bromide	iodide	

CO_2
carbon diox*ide*

$CaBr_2$
calcium brom*ide*

KCl
potassium chlor*ide*

Li_2S
lithium sulf*ide*

NO_3
nitrogen triox*ide*

Prefixes, like *bi-* (for two) and *tri-* (for three), are used for binary covalent compounds formed from two nonmetals. In this case, the prefix matches the subscript. For example, N_2O_3 is read as dinitrogen trioxide. You can find a list of common prefixes in the next section. It's important to realize that we only use the prefixes for a binary covalent compound between two nonmetals. Students who don't realize this often make the mistake of using prefixes when they shouldn't be used. There is a reason that we only use the prefixes for binary covalent bonds, which we will explain in the following section. If you understand this reason, it will help you to remember when to – and not to – use the prefixes.

In an ionic bond between a metal and nonmetal, always write the metal first, as in magnesium oxide (MgO). One exception is the ammonium ion, in which case the bond is ionic, but no metal is present. We still write the positive ion first, as in $(NH_4)_2S$, which we call ammonium sulfide.

When a compound consists of three or more elements, the atom groups have special names. For example, $MgSO_4$ is magnesium sulfate. When you see three or more elements together, you should consult the table of common polyatomic ions on page 50 in order to name the compound correctly.

Ionic compounds involving the transition metals between groups 2 and 3 of the periodic table (see the periodic table on page 16) or the metals in groups 3 thru 6 (but not aluminum) include Roman numerals to clarify the metal ion involved. Unlike the alkali metals, alkaline earth metals, and aluminum, many of the other metals form multiple ions because of their complex subshell structure. For example, iron (Fe) commonly forms both Fe^{2+} and Fe^{3+}. The Roman numeral is included in the name of the compound in order to avoid ambiguity. For example, $FeCl_2$ is called iron (II)

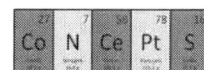

chloride because the Fe^{2+} ion is involved in the bond – not Fe^{3+}. We will discuss this in more detail in Sec. 3.3.

Some compounds involving hydrogen (**H**) have special names, like water (H_2O) and ammonia (NH_3). Acids, such as hydrochloric acid (**HCl**), also begin with hydrogen and follow a different naming scheme. There are also numerous organic compounds involving carbon (**C**), like methane (CH_4). We will consider such special names in Sec.'s 3.4-3.5.

The following flowchart provides a general (though there is an occasional exception) guide for how to name a compound. The first step is to determine whether or not the bond is ionic or covalent.

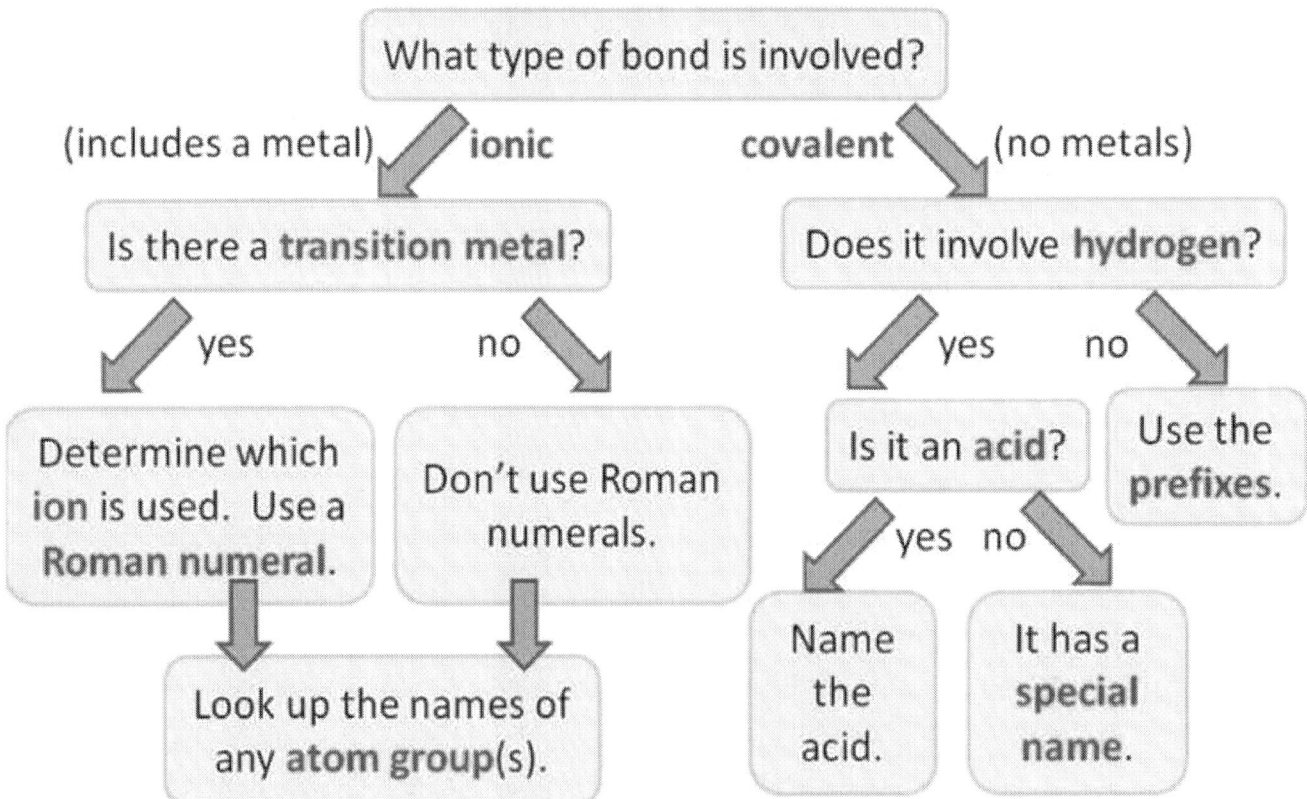

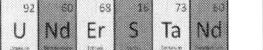 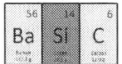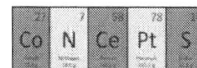
Naming compounds Do's & Don't's

- Do use prefixes for **binary covalent compounds** like CO_2.

- Don't use prefixes (like di- and tri-) for **ionic compounds** like $MgCl_2$.

- Do use Roman numerals if the compound involves a **transition metal** like Fe_2O_3.

- Don't use Roman numerals if the compound does **not** involve a **transition metal** like Al_2O_3.

CO_2
carbon dioxide

$MgCl_2$
magnesium chloride

Fe_2O_3
iron (III) oxide

Al_2O_3
aluminum oxide

3.2 Binary Covalent Bonds

B**i**nary covalent compounds consist of two nonmetals sharing electrons, as in sulfur trioxide (SO_3). As usual in a binary compound, the second element ends with the suffix *-ide*. It may be helpful to consult the table of common endings for binary compounds tabulated in Sec. 3.1.

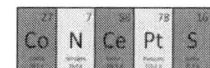

In addition to changing the suffix of the second nonmetal to *-ide*, prefixes like *di-* and *tri-* are also included when naming binary covalent compounds. The prefixes corresponding to the subscripts one thru five can be found in the following table.

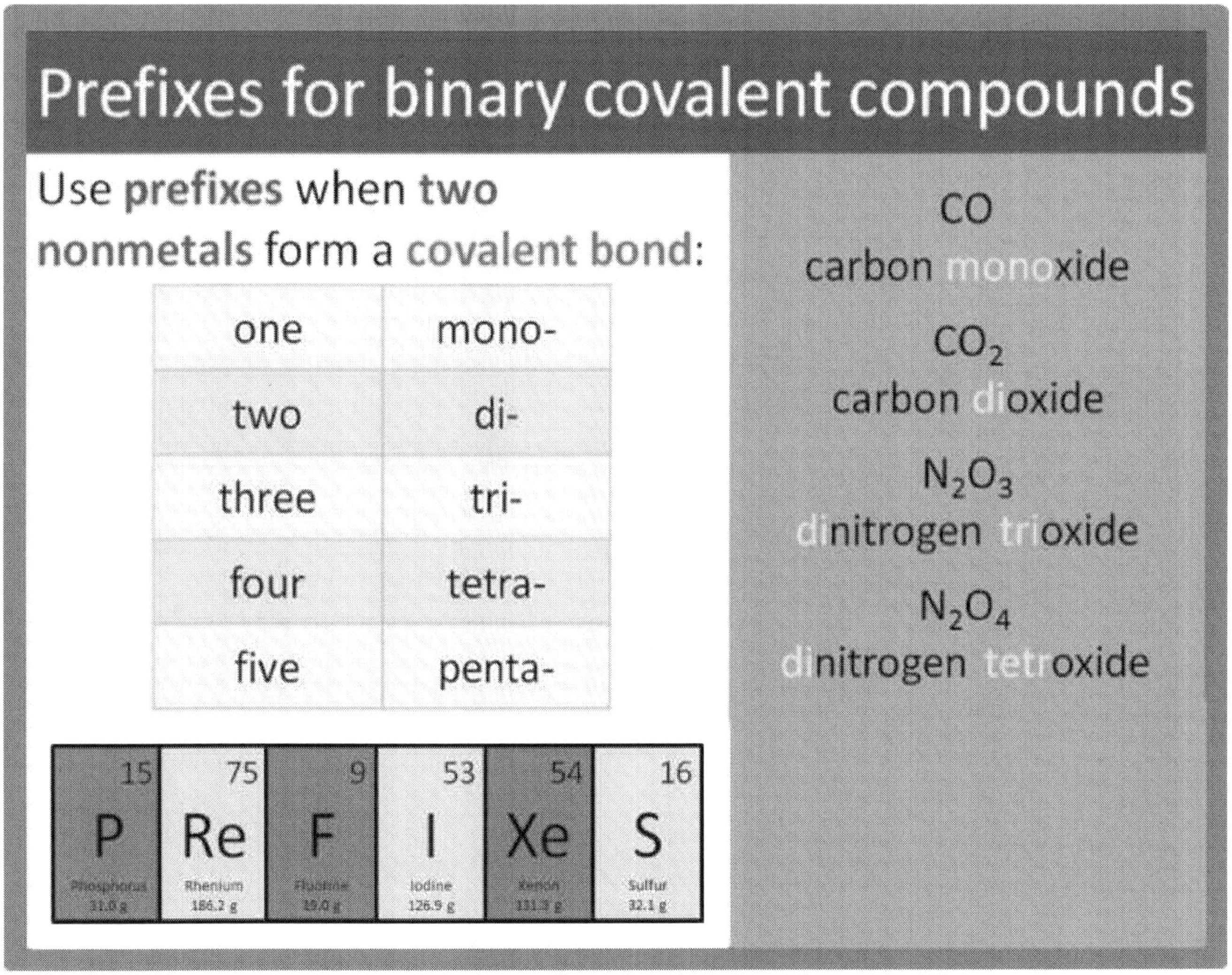

Prefixes for binary covalent compounds

Use **prefixes** when two nonmetals form a **covalent bond**:

one	mono-
two	di-
three	tri-
four	tetra-
five	penta-

15	75	9	53	54	16
P	Re	F	I	Xe	S
Phosphorus 31.0 g	Rhenium 186.2 g	Fluorine 19.0 g	Iodine 126.9 g	Xenon 131.3 g	Sulfur 32.1 g

CO
carbon monoxide

CO_2
carbon dioxide

N_2O_3
dinitrogen trioxide

N_2O_4
dinitrogen tetroxide

When naming binary covalent compounds, include a prefix for each subscript. For example, CO_2 is carbon dioxide and N_2O_3 is dinitrogen trioxide.

Use the prefix *mono-* when the second nonmetal does not have a subscript. For example, CO is carbon monoxide and NO is nitrogen monoxide. Note, however, that we don't include the prefix *mono-* with the first nonmetal.

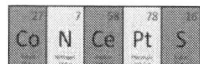

When writing the formula for a covalent compound, the more active nonmetal comes second. Recall the trends for chemical activity from Sec. 1.4, which are illustrated on the periodic table on page 24. In particular, nonmetals are more active going to the right and up in the periodic table (except for the noble gases, which are highly inactive). For example, oxygen difluoride (**OF₂**) has oxygen (**O**) first because fluorine (**F**) is more active chemically than oxygen.

Name the compound NO.

Nitrogen monoxide.

Name the compound ICl₃.

Iodine trichloride.

Name the compound PCl₅.

Phosphorus pentachloride.

Write the chemical formula for dinitrogen tetroxide.

N₂O₄.

Write the chemical formula for phosphorus trichloride.

PCl₃.

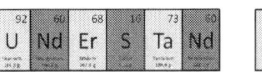 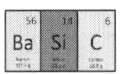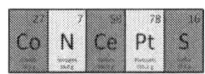
The reason that we include the prefixes when naming binary covalent compounds is that covalent compounds can often share electrons in more than one way. For example, carbon and oxygen can form carbon monoxide (CO) or carbon dioxide (CO_2). Thus, the prefixes are used to avoid ambiguity when naming the compounds.

Do <u>not</u> use the prefixes when naming other types of compounds, like ionic bonds. <u>Only</u> use the prefixes when naming a covalent compound between two elements.

3.3 Ionic Bonds

onic bonds either involve a metal or a positive polyatomic ion, like ammonium (NH_4^+), in combination with a nonmetal or a negative polyatomic ion, like sulfate (SO_4^{2-}). In a binary ionic compound, the second element ends with the suffix -*ide*. It may be helpful to consult the table of common endings for binary compounds tabulated in Sec. 3.1. However, unlike binary covalent compounds, the names for ionic compounds do <u>not</u> involve prefixes.

When the ionic bond involves a metal, the metal always comes first, as in potassium chloride (KCl). Otherwise, the positive polyatomic atom group comes first, as in ammonium sulfide (($NH_4)_2S$).

When there are two or more nonmetals present in an ionic bond, there is at least one polyatomic ion involved in the bond. In this case, you can look up the polyatomic ion in the table in Sec. 2.3.

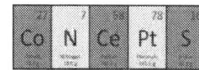

When there is a transition metal (i.e. a metal between groups 2 and 3 – see the periodic table on page 16) or a metal in groups 3 thru 6 and below aluminum (**Al**), include a Roman numeral in parentheses to indicate the charge of the ion in the name of the ionic compound. For example, the copper ion Cu^{2+} combines with oxygen (**O**) to form copper (II) oxide (**CuO**). The Roman numeral (II) corresponds to the charge of the positive ion Cu^{2+}.

The reason for including the Roman numeral with the transition metals and the metals between groups 3 and 6 and below aluminum (**Al**) is that they often form multiple ions due to their rich subshell structure. For example, iron can form Fe^{2+} or Fe^{3+}. It may be helpful to refer to the table in Sec. 2.3 when naming an ionic bond that involves a transition metal or a metal between groups 3 and 6 and below aluminum (**Al**).

In order to name a compound with a transition metal or a metal between groups 3 and 6 and below aluminum (**Al**), first determine which ion is involved in the compound. Determine the ion of the metal the same way that we write the chemical formula for an ionic bond, but backwards. See Sec. 2.3.

For example, consider the chemical formula **FeO**. In order to name this compound correctly, we need to determine which ion of iron (**Fe**) is involved in the ionic bond. In order to figure this out, we're going to use the technique from Sec. 2.3, but we will apply it backwards. The natural ion of oxygen (**O**) is O^{2-}. Therefore, the ion of iron involved in **FeO** must be Fe^{2+}. We can deduce this because the superscripts must cancel when we cross them to make the subscripts in order for the two ions to form **FeO**. Therefore, we name this iron (II) oxide, where the Roman numeral II corresponds to the charge of Fe^{2+}.

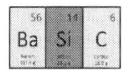 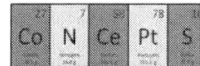

In contrast, consider the chemical formula Fe_2O_3. This time the ion of iron involved must be Fe^{3+} because the superscripts of the ion must make the subscripts of Fe_2O_3 when they are crossed (this technique is the reverse of what we did in Sec. 2.3). Therefore, this compound is called iron (III) oxide. The Roman numeral III refers to the charge of Fe^{3+}.

It is important to note that the Roman numerals are <u>only</u> used for an ionic bond that involves a transition metal or a metal in groups 3 thru 6 that is below aluminum (Al). Do <u>not</u> use a Roman numeral when the metal is an alkali metal (group 1), an alkaline earth metal (group 2), an aluminum (Al) atom, or a positive polyatomic ion like ammonium (NH_4^+). Also, do <u>not</u> use Roman numerals for covalent bonds.

Name the compound KI.

Potassium iodide. No Roman numerals.

Name the compound $CaBr_2$.

Calcium bromide. No Roman numerals.

Name the compound HgS.

Mercury (II) sulfide. Sulfur is S^{2-}, so mercury is Hg^{4+}.

Name the compound FeF_3.

Iron (III) fluoride. Fluorine is F^-, so iron is Fe^{3+}.

Name the compound $Sn(SO_4)_2$.

Tin (IV) sulfate. Sulfate is SO_4^{2-}, so tin is Sn^{2+}.

Given the name of an ionic compound, follow these steps in order to determine the chemical formula:

- Determine the metal and nonmetal ions. For a single nonmetal and for a metal that is aluminum (**Al**), an alkali metal, or an alkaline earth metal, write the natural ion as we did in Sec.'s 2.2-2.3. If the metal is a transition element or a metal in groups 3 thru 6 below aluminum, the positive charge of its ion is indicated by the Roman numeral. For a polyatomic ion, consult the table in Sec. 2.3.
- Next, swap the superscripts to make the subscripts as we did in Sec. 2.3. If there are any 1's, don't write them.
- Check if your answer can be reduced. For example, Ca_2O_2 becomes **CaO**.

Following are a few examples for how to determine the chemical formula of an ionic compound:

- Sodium fluoride. Their natural ions are Na^+ and F^-. Cross the superscripts to make **NaF**.
- Tin (II) oxide. The Roman numeral II tells us that the tin (**Sn**) ion is Sn^{2+}, and the natural ion of oxygen (O) is O^{2-}. Cross the superscripts to make Sn_2O_2, which reduces to **SnO**.
- Copper (II) nitrate. The Roman numeral II tells us that the copper (**Cu**) ion is Cu^{2+}, while the table of polyatomic ions in Sec. 2.3 tells us that the nitrate ion is NO_3^-. Cross the superscripts to make $Cu(NO_3)_2$.

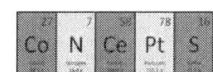

Write the chemical formula for magnesium chloride

1. Mg^{2+}, Cl^-. 2. $MgCl_2$.

Write the chemical formula for sodium sulfide.

1. Na^+, S^{2-}. 2. Na_2S.

Write the chemical formula for lead (IV) chloride.

1. Pb^{4+}, Cl^-. 2. $PbCl_4$.

Write the chemical formula for nickel (II) oxide.

1. Ni^{2+}, O^{2-}. 2. Ni_2O_2. 3. NiO.

Write the chemical formula for aluminum sulfate.

1. Al^{3+}, SO_4^{2-}. 2. $Al_2(SO_4)_3$.

3.4 Special Names

ome compounds have special names. This is the case with many compounds that include hydrogen (**H**) or carbon (**C**).

When hydrogen is present, the compound often has a special name. Many of the compounds that involve water are acids when a solution is made with the compound and water. Acids follow a different naming scheme, which we will discuss in the next section.

A couple of common compounds that involve hydrogen and have special names are water (H_2O) and ammonia (NH_3). Don't make the mistake of calling these dihydrogen monoxide or nitrogen trihydride!

Note the distinction between ammonia and ammonium. Ammonia (NH_3) is a binary covalent compound, whereas ammonium (NH_4^+) is a positive polyatomic ion. Ammonia appears by itself, while ammonium forms ionic compounds like ammonium chloride (NH_4Cl).

One way to remember which formula matches which name is to understand the covalent bond between nitrogen (N) and hydrogen (H). Nitrogen needs 3 electrons to fill its outer shell (since it has 5 valence electrons in an outer shell that can hold up to 8), while hydrogen needs 1 electron. Therefore, NH_3 is a neutral compound because the 3 hydrogen atoms supply the 3 valence electrons that nitrogen needs by sharing a pair of electrons with nitrogen, which also supplies the 1 valence electron that each hydrogen needs. In contrast, a bond between 1 nitrogen and 4 hydrogens has one too many electrons, so it tends to form the NH_4^+ ion.

Another way to remember which formula goes with ammonia and which goes with ammonium is to realize that ammonia comes before ammonium alphabetically, which matches that NH_3 comes before NH_4^+ numerically (i.e. the 3 comes before the 4).

Carbon (C) forms numerous compounds, most of which have special names, like methane (CH_4). The wide variety of complex molecules that make up living organisms contain carbon. The chemistry of carbon is very rich – and very significant for biology. There is actually an entire branch devoted to the chemistry of carbon called organic chemistry. Organic molecules are molecules that contain one or more carbon atoms. Inorganic

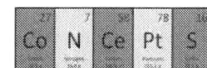

molecules do not contain any carbon atoms. However, two exceptions are carbon monoxide (CO) and carbon dioxide (CO_2), which are inorganic.

Organic compounds don't follow the usual naming scheme. The names and chemical formulas for some common organic compounds include benzene (C_6H_6), butane (C_4H_{10}), glucose ($C_6H_{12}O_6$), methane (CH_4), octane (C_8H_{18}), and propane (C_3H_8). Some of these formulas look reducible, like C_6H_6. However, C_6H_6, for example, is not reducible because the molecule actually has 6 carbon atoms and 6 hydrogen atoms bonded together.

Name the compound K_2O.

Potassium oxide. No prefixes, no Roman numerals.

Name the compound NH_3.

Ammonia. This compound has a special name.

Name the compound PbF_2.

Lead (II) fluoride. The lead ion is Pb^{2+}.

Name the compound N_2O.

Dinitrogen monoxide. Binary covalent: Use prefixes.

Name the compound HCl.

Hydrochloric acid. It begins with H (and it's not water).

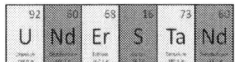 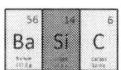 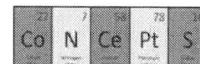

3.5 Acids

...ids also follow a different naming scheme. Acids are generally compounds that begin with hydrogen – except for water (H_2O), which is not an acid. The term acid usually refers to a solution of the compound in water, rather than referring to the compound itself.

Some common acids include hydrochloric acid (**HCl**), nitric acid (**HNO_3**), phosphoric acid (**H_3PO_4**), and sulfuric acid (**H_2SO_4**). We will discuss acids and bases in Sec. 5.7.

Write the chemical formula for carbon dioxide.

CO_2. Binary covalent compounds use prefixes.

Write the chemical formula for methane.

CH_4. This is one of the special names.

Write the chemical formula for magnesium bromide.

$MgBr_2$. The ions are Mg^{2+} and Br^-.

Write the chemical formula for potassium sulfide.

K_2S. The ions are K^+ and S^{2-}.

Write the chemical formula for nickel (II) fluoride.

NiF_2. The ions are Ni^{2+} and F^-.

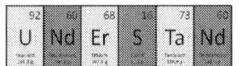 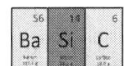 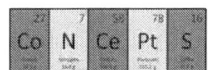

Ch. 3 Self-Check Exercises

1. Write out the names for the following compounds:

LiF	CO$_2$	NH$_3$	Fr$_2$O	FeCl$_3$
NaOH	CuSO$_4$	BaCl$_2$	N$_2$O$_3$	NaCl
Sn(NO$_3$)$_2$	H$_2$SO$_4$	PCl$_3$	MgF$_2$	(NH$_4$)$_2$SO$_3$

2. Write the chemical formulas for the following compounds:

carbon monoxide	water	calcium chloride
iron (II) bromide	dinitrogen tetroxide	sodium oxide
beryllium fluoride	tin (II) chloride	dinitrogen monoxide
sodium nitrate	ammonium oxide	iron (III) sulfate

4. F Chemical Reactions

4.1 Counting Atoms

fore we discuss chemical reactions, let us discuss the significance of a chemical formula and how to count the atoms of a given element in a single term of a chemical reaction. This will help to understand what it means for the equation for a chemical reaction to be balanced, which we will discuss in the following section. In Sec. 4.3, we will describe other concepts associated with chemical reactions, while in the first two sections we will focus on the concepts associated with counting atoms and interpreting the symbolic equation for a chemical reaction.

The significance of a chemical formula is that it tells you how many of which type of atom are bonded together. It also tells you the nature of the bond – i.e. whether or not the bond is ionic or covalent.

For example, each ammonia (NH_3) molecule has 3 hydrogen (**H**) atoms bonded to 1 nitrogen (**N**) atom. The elements in the chemical formula tell us what the substance is composed of, and the subscripts tell us how many of each type of atom are involved in the bond. The fact that this is a bond between two nonmetal atoms tells us that the nitrogen and hydrogen atoms share electrons – i.e. the bond is covalent.

As another example, iron (III) oxide (Fe_2O_3) consists of ionic bonds – since iron (**Fe**) is a metal – with 2 Fe^{3+} ions attracted to 3 O^{2-} ions. We also know that both iron atoms donated 3 electrons, and each oxygen (**O**) atom received 2 electrons.

A subscript denotes when more than one atom of a given element is present in a molecule, as illustrated by the following examples:

One **H₂O** (water) molecule consists of 2 hydrogen (**H**) atoms and 1 oxygen (**O**) atom.

One **C₃H₈** (propane) molecule consists of 3 carbon (**C**) atoms and 8 hydrogen (**H**) atoms.

One **K₂CO₃** (potassium carbonate) molecule consists of 2 potassium (**K**) atoms, 1 carbon (**C**) atom, and 3 oxygen (**O**) atoms.

One **NH₄NO₃** (ammonium nitrate) molecule consists of 2 nitrogen (**N**) atoms, 4 hydrogen (**H**) atoms, and 3 oxygen (**O**) atoms.

A subscript that follows parentheses () multiplies the numbers of all of the atoms inside the parentheses, as illustrated by the following examples:

Pb(NO₃)₂ (lead (II) nitrate) consists of 1 lead (**Pb**) atom, $2 \times 1 = 2$ nitrogen (**N**) atoms, and $2 \times 3 = 6$ oxygen (**O**) atoms.

Ca(OH)₂ (calcium hydroxide) consists of 1 calcium (**Ca**) atom, 2 oxygen (**O**) atoms, and 2 hydrogen (**H**) atoms.

Mg₃(PO₄)₂ (magnesium phosphate) consists of 3 magnesium (**Mg**) atoms, 2 phosphorus (**P**) atoms, and 8 oxygen (**O**) atoms.

Al₂(SeO₄)₃ (aluminum selenate) consists of 2 aluminum (**Al**) atoms, 3 selenium (**Se**) atoms, and 12 oxygen (**O**) atoms.

A coefficient is a number that appears to the left of a molecule. The coefficient multiplies the numbers of all atoms in the molecule, as illustrated by the following examples:

- $4H_2O$ is 4 water molecules: $4 \times 2 = 8$ hydrogen (**H**) atoms and $4 \times 1 = 4$ oxygen (**O**) atoms.
- $5Fe_2O_3$ (iron (III) oxide) consists of 10 iron (**Fe**) atoms and 15 oxygen (**O**) atoms.
- $6N_2$ (diatomic nitrogen gas) consists of 12 nitrogen (**N**) atoms.
- $3NiSO_4$ (nickel (II) sulfate) consists of 3 nickel (**Ni**) atoms, 3 sulfur (**S**) atoms, and 12 oxygen (**O**) atoms.
- $8C_2H_5OH$ (ethanol) consists of 16 carbon (**C**) atoms, 48 hydrogen (**H**) atoms, and 8 oxygen (**O**) atoms.

A coefficient and a subscript after parentheses () both multiply the numbers of all of the atoms inside the parentheses, as illustrated by the following examples:

- $4Al_2(SO_4)_3$ (aluminum sulfate) consists of $4 \times 2 = 8$ aluminum (**Al**) atoms, $4 \times 3 = 12$ sulfur (**S**) atoms, and $4 \times 3 \times 4 = 48$ oxygen (**O**) atoms.
- $2Ba(ClO_3)_2$ (barium chlorate) consists of 2 barium (**Ba**) atoms, 4 chlorine (**Cl**) atoms, and 12 oxygen (**O**) atoms.
- $3(NH_4)_2SO_4$ (ammonium sulfate) consists of 6 nitrogen (**N**) atoms, 24 hydrogen (**H**) atoms, 3 sulfur (**S**) atoms, and 12 oxygen (**O**) atoms.

 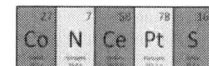
4.2 Balancing Equations

 hemical reactions occur when two or more substances undergo mutual chemical changes. The initial compounds are called the reactants (because they react together) and the final compounds are called the products (because they are what the reaction produced).

A chemical equation indicates which compounds are involved in the reaction. The left-hand side of the equation has the reactants added together, while the right-hand side of the equation has the products added together. The "equation" actually has a yield ($\rightarrow$) symbol rather than an equal sign ($=$). The reaction "yields" the products. An example of a chemical reaction is $2H_2 + O_2 \rightarrow 2H_2O$. In this example, the reactants – diatomic hydrogen gas (H_2) and diatomic oxygen gas (O_2) – react together to produce water vapor (H_2O). The coefficients indicate how much of each compound are involved in the reaction.

A balanced equation tells you how much of each reactant is needed to form the products. The equation is balanced if there are equal numbers of each type of atom on both sides.

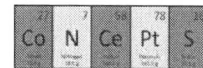

For example, consider the equation for the formation of water: $2H_2 + O_2 \rightarrow 2H_2O$. This equation is balanced because there are 4 hydrogen (H) atoms and 2 oxygen (O) atoms on both sides of the equation. The coefficients (i.e. the 2's) before H_2 and H_2O are needed to balance the equation. That is, we couldn't write $H_2 + O_2$ on the left-hand side and H_2O on the right-hand side because there would then be 2 O atoms on the left, but 1 O atom on the right.

Sometimes, it is necessary to balance a chemical equation. For example, suppose that you know that zinc (Zn) is reacting with sulfur (S) in the form S_8 to yield zinc (II) sulfide (ZnS). In this case, you know that the structure of the reaction looks like $Zn + S_8 \rightarrow ZnS$. However, this equation is not balanced: There are 8 S atoms on the left, but only 1 on the right.

We balance a chemical reaction by adding coefficients until there are equal numbers of each type of atom on both sides of the equation. First we add a coefficient 8 to the right-hand side of the equation in order to balance the S atoms: $Zn + S_8 \rightarrow 8ZnS$. This balances S, but Zn is no longer balanced. Finally, we add another coefficient 8 to the left-hand side in order to balance Zn: $8Zn + S_8 \rightarrow 8ZnS$. Now both Zn and S are balanced: There are 8 Zn and 8 S atoms on both sides of the equation.

When balancing a chemical equation, the goal is to have the same number of each type of atom on each side of the equation. When you think that the equation is balanced, count the number of each type of atom on each side of the equation and see if it is, in fact, balanced. If the equation is not balanced, add a coefficient somewhere to try and balance one type of atom. Sometimes, when you add a coefficient, it balances one type of atom, but a different atom becomes unbalanced. For this reason, you must use some trial and error, and logic and reason – and it may take a few steps before the equation is finally balanced.

27	7	16	68	23	85	53	8	7
Co	N	S	Er	V	At	I	O	N
Cobalt 58.9 g	Nitrogen 14.0 g	Sulfur 32.1 g	Erbium 167.3 g	Vanadium 50.9 g	Astatine 210.0 g	Iodine 126.9 g	Oxygen 16.0 g	Nitrogen 14.0 g

91	32
Pa	Ge
Protactinium 231.0 g	Germanium 72.6 g

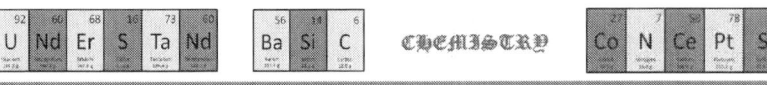

The following rules are generally useful, but bear in mind that there are a few exceptions to these rules:

- Begin by balancing one element at a time.
- First balance elements that appear only once on each side of the equation. Balance multi-element compounds like **NaCl** before balancing single-element terms like N_2. Balance **H** and **O** atoms last.
- Use trial and error. Be patient.
- Add up all of the kinds of atoms on both sides of the equation to make sure that it's completely balanced.

The following examples help to illustrate the technique of balancing a chemical reaction.

- $Fe_2O_3 + CO \rightarrow Fe + CO_2$. First add a coefficient to balance iron (**Fe**): $Fe_2O_3 + CO \rightarrow 2Fe + CO_2$. Carbon (**C**) is already balanced, so last is oxygen (**O**). We can balance oxygen using two 3's: $Fe_2O_3 + 3CO \rightarrow 2\,Fe + 3CO_2$. Check: We have 2 **Fe**, 3 **C**, and 6 **O** atoms on both sides of the equation.
- $ZnS + O_2 \rightarrow ZnO + SO_2$. Here, zinc (**Zn**) and sulfur (**S**) are already balanced, but oxygen (**O**) is not: There are 2 **O** atoms on the left, but 3 on the right. If we add a coefficient 2 to the O_2 term, there will be 4 on the left, but 3 on the right. We could then add a 2 to the **ZnO** term to balance **O**, but then **Zn** won't be balanced. Instead, consider this idea, which often comes in handy: When there are 3 on one side and 2 on the other, try multiplying 3 times 2 and 2 times 3, which both make 6. Here, we do that as follows: $ZnS + 3O_2 \rightarrow 2ZnO + 2SO_2$. Just add one last coefficient to balance **Zn** and **S**: $2ZnS + 3O_2 \rightarrow 2ZnO + 2SO_2$. Now there are 2 **Zn**, 2 **S**, and 6 **O** atoms on both sides of the equation, so the reaction has been balanced.
- $C_3H_8 + O_2 \rightarrow CO_2 + H_2O$. Putting a coefficient 4 with H_2O will balance hydrogen (**H**), while inserting a 3 with CO_2 will balance carbon

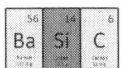 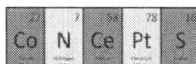

(C): $C_3H_8 + O_2 \rightarrow 3CO_2 + 4H_2O$. Now there are 10 oxygen (O) atoms on the right, so we need a coefficient 5 with O_2: $C_3H_8 + 5O_2 \rightarrow 3CO_2 + 4H_2O$. This is equation is now balanced, with 3 **C**, 8 **H**, and 10 **O** atoms on both sides.

Balance the equation $NO + O_2 \rightarrow NO_2$.

$2NO + O_2 \rightarrow 2NO_2$.

Balance the equation $Fe + Cl_2 \rightarrow FeCl_3$.

$2Fe + 3Cl_2 \rightarrow 2FeCl_3$.

Balance the equation $C + H_2 \rightarrow C_5H_{12}$.

$5C + 6H_2 \rightarrow C_5H_{12}$.

Balance the equation $Fe + H_2O \rightarrow Fe_3O_4 + H_2$.

$3Fe + 4H_2O \rightarrow Fe_3O_4 + 4H_2$.

Balance the equation $C_6H_{14} + O_2 \rightarrow CO_2 + H_2O$.

$2C_6H_{14} + 19O_2 \rightarrow 12CO_2 + 14H_2O$.

 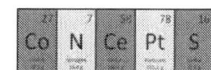

4.3 Types of Reactions

 hemical reactions come in a few basic types, including synthesis, decomposition, and single- and double-replacement. A chemical reaction can also be classified as endothermic or exothermic. In this section, we will explore these kinds of reactions, in addition to a class of reactions called oxidation-reduction reactions.

A synthesis reaction occurs when two or more substances combine together chemically to form a new substance. A synthesis reaction has the following structure: $A + B \rightarrow AB$, where A and B represent two different substances that combined together to form a new substance, AB. One way that a synthesis reaction occurs is when two elements combine together to form a compound. Examples of synthesis include sodium (Na) and diatomic chlorine gas (Cl_2) bonding together ionically to form sodium chloride ($NaCl$), $2Na + Cl_2 \rightarrow 2NaCl$, and carbon ($C$) and diatomic oxygen gas (O_2) bonding together covalently to form carbon dioxide (CO_2), $C + O_2 \rightarrow CO_2$. A synthesis reaction can be a little different than merely bonding two elements together, provided that it still has the same structure as $A + B \rightarrow AB$. For example, the formation of calcium sulfite ($CaSO_3$) from calcium oxide (CaO) and sulfur dioxide (SO_2) is a synthesis reaction: $CaO + SO_2 \rightarrow CaSO_3$.

In a decomposition reaction, a single substance yields two or more simpler substances through chemical change. Decomposition is essentially the reverse of synthesis. A decomposition reaction has the form $AB \rightarrow A + B$. When an electric current is applied to decompose the substance, the decomposition reaction is called electrolysis. Examples of decomposition include sulfur trioxide (SO_3) producing sulfur dioxide (SO_2) and diatomic oxygen gas (O_2), $2SO_3 \rightarrow 2SO_2 + O_2$, and the electrolysis of water (H_2O) into diatomic hydrogen gas (H_2) and diatomic oxygen gas (O_2), $2H_2O \rightarrow 2H_2 + O_2$.

In a single-replacement reaction, one substance replaces another in a compound. A single-replacement reaction has a structure similar to $A + BC \rightarrow B + AC$, where substances A and B swapped roles. An example of single-replacement is $Cl_2 + 2KBr \rightarrow 2KCl + Br_2$: Bromine ($Br$) and chlorine ($Cl$) exchanged their roles in the compounds potassium bromide (KBr) and potassium chloride (KCl). Double-replacement involves swapping the roles of two substances. A double-replacement reaction has the form $AC + BD \rightarrow BC + AD$. For example, $HCl + NaOH \rightarrow NaCl + H_2O$ is a double-replacement reaction.

Any type of reaction may be classified as endothermic or exothermic. The distinction is that endothermic reactions absorb energy, whereas exothermic reactions release energy. Both the reactants and products have stored chemical energy. If the products have more stored energy than the reactants, then energy is absorbed – i.e. the reaction is endothermic. If the reactants have more stored energy than the products, then energy is released – i.e. the reaction is exothermic. The transfer of energy that occurs – i.e. the absorption or release of energy – during the reaction comes in the form of heat. The system absorbs heat in an endothermic reaction and gives off heat in an exothermic reaction.

For example, the formation of water, $2H_2 + O_2 \rightarrow 2H_2O$, is exothermic: The reactants – two parts diatomic hydrogen gas (H_2) and one part diatomic oxygen gas (O_2) – have more stored chemical energy than the product – two parts water (H_2O). Heat is released when H_2 and O_2 combine

to form H_2O. In contrast, the reverse process, the electrolysis of water, $2H_2O \rightarrow 2H_2 + O_2$, is endothermic: The products – two parts H_2 and one part O_2 – have more stored energy than the reactant – two parts H_2O. Heat is absorbed when H_2O decomposes into H_2 and O_2. The reason that heat is released in one reaction, but absorbed in the reverse reaction, has to do with conservation of energy (which we will discuss in the next section).

Oxidation-reduction (abbreviated redox) reactions are chemical reactions that involve a transfer of electrons. For example, the synthesis of sodium chloride ($NaCl$) from sodium (Na) and diatomic chlorine gas (Cl_2), $2Na + Cl_2 \rightarrow 2NaCl$, is a redox reaction. A redox reaction is actually a combination of two half-reactions – an oxidation half-reaction and a reduction half-reaction. In this example, Na is oxidized in the half reaction $Na \rightarrow Na^+ + e^-$, while Cl is reduced in the half-reaction $Cl_2 + 2e^- \rightarrow 2Cl^-$. Each Na loses an electron, whereas each Cl gains an electron.

During oxidation, an atom loses electrons, while during reduction, an atom gains electrons. The definition of reduction as a gain in electrons usually seems counterintuitive to students. Think about it: The atom is 'reduced' when it 'gains' electrons. Does that seem strange to you? Actually, the atom is 'reduced' in the sense that it becomes negatively charged after gaining electrons. Its charge is what is reduced in the sense of becoming negatively charged.

Many important redox reactions involve oxygen (O). Hence the term oxidation. When oxygen is involved in the redox reaction, oxidation occurs when a substance combines with oxygen and reduction occurs when oxygen is removed from a substance. The substance that gains oxygen is oxidized; the substance that loses oxygen is reduced. For example, in the reaction

$2\mathbf{Al} + \mathbf{Fe_2O_3} \rightarrow \mathbf{Al_2O_3} + 2\mathbf{Fe}$, aluminum (**Al**) is oxidized (it gains oxygen) and iron (**Fe**) is reduced (it loses oxygen).

Combustion is a redox reaction in which a substance combines with oxygen, where much heat and light are released during rapid oxidation. This occurs, for example, when a substance like wood or gasoline burns, as in the combustion of methane ($\mathbf{CH_4}$): $\mathbf{CH_4} + 2\mathbf{O_2} \rightarrow \mathbf{CO_2} + 2\mathbf{H_2O}$. The synthesis of water ($\mathbf{H_2O}$), $2\mathbf{H_2} + \mathbf{O_2} \rightarrow 2\mathbf{H_2O}$, is another type of combustion reaction. In each case, a substance combines with oxygen and the reaction releases much heat and light.

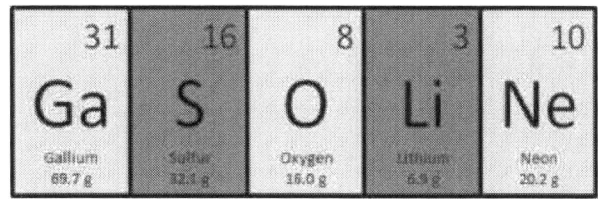

Redox reactions are also important in chemical reactions in electrochemistry. For example, the exchange of electrons involved in redox reactions can be used to make a battery (electrochemical cell). The two substances that gain or lose electrons serve as the electrodes – i.e. the positive and negative terminals. The electrodes are immersed in solutions, which allows the two half-reactions to occur. The substance being oxidized (losing electrons) serves as the negative electrode (anode), while the substance being reduced (gaining electrons) serves as the positive electrode (cathode). For example, in the Daniell cell, zinc (**Zn**) is oxidized, $\mathbf{Zn} \rightarrow \mathbf{Zn^{2+}} + 2\mathbf{e^-}$, at the anode and copper (**Cu**) is reduced, $\mathbf{Cu^{2+}} + 2\mathbf{e^-} \rightarrow \mathbf{Cu}$, at the cathode. When the battery is connected in a circuit, electrons in the conducting wires flow from the negative terminal to the positive terminal. This flow of electricity is what we call current.

When a reactant or product is dissolved in water, we refer to the solution as an aqueous solution. An (aq) often follows the reactant or product to indicate this. It is also common to include (s) for solid, (l) for liquid, and (g) for gas to indicate which state the substance is in. For example, the full reaction for the Daniell cell is $\mathbf{Zn}$ (s) $+ \mathbf{Cu^{2+}}$ (aq) $\rightarrow \mathbf{Zn^{2+}}$

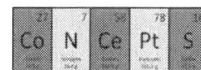
(aq) $+$ **Cu** (s).   The solid zinc (**Zn**) and copper (**Cu**) refer to the metal electrodes, while the Cu^{2+} (aq) and Zn^{2+} (aq) refer to ions in aqueous solutions. In this book, we opted to supress the ()'s in the reactions in order to avoid clutter, since our focus is on understanding the main concepts.

Salts and many other electrolytes dissociate into ions when dissolved in water (H_2O). For example, sodium chloride (**NaCl**) dissociates into Na^+ and Cl^- ions when it dissolves in water. Acids dissociate into H^+ and nonmetal ions, while bases dissociate into OH^- ions and metal ions. For example, the dissociation of hydrochloric acid (**HCl**) is $HCl \rightarrow H^+ + Cl^-$, and the dissociation of sodium hydroxide (**NaOH**), a base, is $NaOH \rightarrow Na^+ + OH^-$. Acids and bases neutralize one another, and form a salt and water in the process, when mixed together.   For example, the neutralization of hydrochloric acid and sodium hydroxide is $HCl + NaOH \rightarrow H_2O + NaCl$. We will consider solutions, including acids and bases, in more detail in Chapter 5.

4.4 Some Common Reactions

Are we will discuss a few chemical reactions that you can relate to everyday experience. We will explore the rusting of iron (corrosion), the chemistry of fire (a type of combustion), how animals obtain their energy (specifically, through the oxidation of glucose), how plants store energy from sunlight (photosynthesis), and why the sun shines (nuclear fusion).

The rusting (or corrosion) of iron (Fe) is a redox reaction. Iron reacts with diatomic oxygen gas (O_2) and water (H_2O) to form iron (III) hydroxide ($Fe(OH)_3$): $4Fe + 3O_2 + 6H_2O \rightarrow 4Fe(OH)_3$. Here, iron is oxidized – it gains oxygen (O) through the hydroxide (OH^-) ion. Rust ($Fe(OH)_3$) is a new substance, distinctly different from iron (Fe), with a reddish brown appearance. Rust involves slow oxidation, whereas combustion, which we will discuss next, involves rapid oxidation.

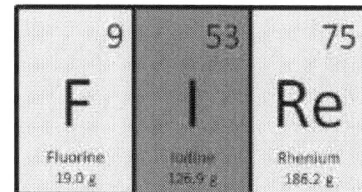

Recall that combustion is a redox reaction in which a substance combines with oxygen (O), where much heat and light is released during rapid oxidation. Combustion occurs, for example, when gasoline, wood, natural gas, or propane burns. Hence, combustion is very important for fire, transportation, and heating. For example, the reaction for the combustion of propane (C_3H_8) is $C_3H_8 + 5O_2 \rightarrow 3CO_2 + 4H_2O$. Combustion also comes in other forms. For example, the synthesis of water (H_2O) from diatomic hydrogen gas (H_2) and diatomic oxygen gas (O_2), $2H_2 + O_2 \rightarrow 2H_2O$, is a combustion reaction because hydrogen combined with oxygen and much heat and light is released during the reaction. Fire was originally believed to involve a hypothetical substance called phlogiston, but it is now understood that fire is really just rapid oxidation (combustion); it turns out that there is no such thing as phlogiston.

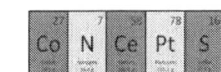

One of the most important reactions involving oxygen for life on earth is photosynthesis. In this reaction, plants combine water (H_2O) and carbon dioxide (CO_2) to form carbohydrates, such as glucose ($C_6H_{12}O_6$), and diatomic oxygen gas (O_2): $6CO_2 + 6H_2O \rightarrow C_6H_{12}O_6 + 6O_2$. The reverse reaction, the oxidation of glucose, $C_6H_{12}O_6 + 6O_2 \rightarrow 6CO_2 + 6H_2O$, occurs when animals eat plants. Food is oxidized, as in the case of digesting glucose, from the oxygen that animals breathe through their lungs. Photosynthesis is endothermic (energy is absorbed by chlorophyll), while the oxidation of glucose is exothermic (that's how animals get their energy by digesting carbohydrates).

The most critical reaction for life on earth occurs in the core of the sun. It is actually a nuclear reaction, rather than chemical reaction, in that it involves a change in the identity of the atoms: Every second, the sun converts four million tons of mass into energy by "fusing" hydrogen (H) nuclei into helium (He) nuclei. Specifically, 4 hydrogen nuclei bind together to form a helium nucleus in an exothermic reaction: $4H \rightarrow He$. This process is called nuclear fusion. This nuclear energy is the energy that causes the sun to shine. The difference in mass between 4 hydrogen nuclei and 1 helium nucleus causes energy to be released in nuclear fusion through Einstein's famous equation, $E = mc^2$, which expresses a relationship between mass and energy (c is the speed of light, which is about 300,000,000 m/s) – see the next section.

4.5 Properties of Reactions

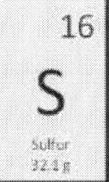 o far, we have focused on quantitative aspects of chemical reactions – the regrouping of elements from the reactants to the products, and how to balance chemical equations with coefficients. We will now consider some qualitative properties of chemical reactions.

When a chemical reaction occurs, there is a change in chemical identity in the substances. The products are chemically different from the reactants. If you could look microscopically at the molecules of the reactants and products, you would see this change in chemical configuration. For example, in the synthesis of water, $2H_2 + O_2 \rightarrow 2H_2O$, initially you would see pairs of hydrogen (**H**) atoms bound together (**H_2**) and pairs of oxygen (**O**) atoms bound together (**O_2**), and later you would see water molecules with 2 **H** atoms and 1 **O** atom bound together. That is, you would see different chemical bonds in the reactants and products.

However, when you "see" a chemical reaction occur, your eye doesn't get a microscopic look at how the atoms are bonding together in the molecules. Instead, you see macroscopic changes. Thus, to your eye, indications that a chemical reaction has occurred are much more subtle. Following are some qualitative indications of a chemical reaction – indications that a chemical (rather than physical) change has occurred. Bear in mind that one of these indications all by itself does not provide evidence of a chemical reaction – e.g. some physical reactions produce heat and light or result in a gas – but a combination of these indications usually does signify that a chemical change has occurred.

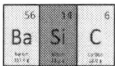 

A chemical reaction that is exothermic may show this release of energy in the form of heat and light. For example, when cooking, the gas from the stove reacts with oxygen (**O**) in the air in combustion, producing much heat and light.

If one of the products of a chemical reaction is a gas, the product may appear in the form of gas bubbles. For example, when baking soda is mixed with vinegar, bubbles of carbon dioxide (CO_2) are produced.

If two solutions are mixed together and one of the products is a solid, this is generally evidence of a chemical reaction. We call the solid a precipitate when a solid forms from the mixture of two solutions and separates from the solution. A precipitate forms when the ions of two solutions separate during the mixture. For example, when solutions of barium chloride ($BaCl_2$) and sodium sulfate (Na_2SO_4) are mixed together, the barium (Ba^{2+}) and sulfate (SO_4^{2-}) ions form barium sulfate ($BaSO_4$), which is a solid.

When a change in color is observed, this is often a sign that a chemical reaction has occurred. For example, when hydrogen iodide (**HI**), a colorless gas, decomposes into diatomic hydrogen gas (H_2) and diatomic iodine vapor (I_2), the iodine vapor that forms appears violet in color.

Let's consider some properties of chemical reactions that you could measure. One quantity that you could measure is energy. If the reactants have more energy than the products, the reaction is exothermic (it releases energy into the surroundings); and if the reactants have less energy than the products, the reaction is endothermic (it absorbs energy from the surroundings). The difference in energy generally manifests itself in the form of heat. The net heat flow is from the system to the surroundings in an

exothermic reaction, and from the surroundings to the system in an endothermic reaction.

Energy is conserved in nature. This means that the total energy of the system plus the surroundings remains constant. Energy may be transformed from one form to another, but cannot be created or destroyed. In a chemical reaction, energy from the reactants is transferred to the products and energy is transferred between the system and the surroundings in the form of heat. In general, we can express conservation of energy for a chemical reaction as:

$$E_{reactants} + Q \rightarrow E_{products}$$

The heat exchanged between the system and surroundings, denoted by the symbol Q, is positive for an endothermic reaction and negative for an exothermic reaction.

In a chemistry course, it is common to work with a quantity called enthalpy, denoted by the symbol H. The change in enthalpy equals the amount of energy absorbed or lost by the system in the form of heat during a reaction that occurs at constant pressure. So as long as the pressure doesn't change during the chemical reaction, the change in enthalpy is synonymous with the heat exchanged.

Technically, mass is not conserved, even though many chemistry textbooks speak of conservation of mass. Unlike energy, there is no fundamental law of nature regarding conservation of mass. When textbooks mention conservation of mass, what they are really referring to is a conservation of the identity of the atoms during a chemical reaction. That is, we have the same number of each type of atom on both sides of a balanced chemical equation. All chemical reactions are balanced. It is a fundamental law of chemistry that you can't gain or lose any type of atom in a chemical reaction. The same elements are involved in both sides of a chemical equation; they simply regroup themselves by changing the chemical bonds. For example, in the chemical reaction $Fe_2O_3 + 3CO \rightarrow 2Fe + 3CO_2$, there are 2 iron (**Fe**), 6 (**O**), and 3 (**C**) atoms on both sides of the equation. Iron atoms are still iron atoms, for example – they didn't change into other types of atoms, like oxygen or carbon. What happened during this reaction is that the oxygen atoms that were bonding with iron are now bonding with carbon, along with the other oxygen atoms that had already been bonding with carbon (and they are now bonding with carbon in a different way, forming a new substance – carbon dioxide rather than carbon monoxide).

It is a fundamental conservation law that the number and type of each atom is conserved during a chemical reaction. For example, if you have 5 moles of oxygen (**O**) in the reactants, you will find 5 moles of oxygen in the products. This is sometimes called conservation of mass, but it is really conservation of the number of each type of atom. This is not a fundamental conservation law of nature, however – it only applies to chemical reactions. In a nuclear reaction, the atoms can actually change from one kind to another, as in nuclear fusion, $4H \rightarrow He$, which occurs in the sun's core. Here, hydrogen (**H**) nuclei collide together at extremely high temperature and pressure to form helium (**He**) nuclei. Energy is conserved in nature, but mass and the number of each type of atom are not conserved in general.

The distinction between mass and energy has to do with Einstein's famous equation, $E = mc^2$, where c is the speed of light, which is about three hundred million meters per second (300,000,000 m/s). This equation expresses a relationship between mass and energy.[1]

75	57	22	8	7	16	1	53	15
Re	La	Ti	O	N	S	H	I	P
Rhenium	Lanthanum	Titanium	Oxygen	Nitrogen	Sulfur	Hydrogen	Iodine	Phosphorus
186.2 g	138.9 g	47.9 g	16.0 g	14.0 g	32.1 g	1.0 g	126.9 g	31.0 g

---

[1] If you've studied Einstein's theory of relativity, you will want to note that we will only use mass to refer to rest-mass, and not relativistic mass.

91	32
Pa	Ge

The mass of a chemical compound is not the same as the sum of the masses of the atoms that make up the compound. The energy that binds the atoms together – called binding energy – affects the mass of the compound through the equation $E = mc^2$. For example, one water (H_2O) molecule consists of 2 hydrogen (H) atoms and 1 oxygen (O) atom. The mass of one H_2O molecule does not equal the mass of 2 H atoms plus the mass of 1 O atom, but the difference in mass does equate to the energy that binds the 2 H atoms and 1 O atom together to make 1 H_2O molecule. When we compute the mass difference and multiply it by the speed of light squared, we find that it equals the binding energy for one water molecule.

Since the sum of the masses of the reactants generally differs from the sum of the masses of the products, mass is not conserved in general. However, energy is conserved, and the difference in mass equates to the energy that binds atoms together to form compounds according to $E = mc^2$.

In the sun's core, hydrogen (H) nuclei collide together at extremely high temperature and pressure to form helium (He) nuclei in a nuclear reaction called nuclear fusion. Unlike a chemical reaction, there is a change in the identity of the atoms, as 4 hydrogen nuclei form 1 helium nucleus: $4H \rightarrow He$.[2] The helium nucleus has slightly less mass than four hydrogen nuclei. As a result, a little energy is released during the reaction according to the equation $E = mc^2$, where m is the difference in mass between 4 hydrogen nuclei and 1 helium nucleus. The sun has so much hydrogen that it converts four million (4,000,000) tons of mass into energy this way every second. This is what causes the sun to shine.

Another quantity that is conserved in chemical reactions is electric charge. Conservation of electric charge is a fundamental conservation law of nature, like conservation of energy. For example, in $Cu^{2+} + 2e^- \rightarrow Cu$, the Cu^{2+} ion is positively charged, the 2 electrons (e^-) are negatively charged, and both sides of the equation are neutral. The charge of an ion is based on the number of protons and electrons in the atom. In chemical reactions, the number of protons and electrons initially and finally do not change. In other

---

[2] The complete reaction also involves the release of electrons, in order to balance electric charge, and other neutral particles.

types of reactions, these numbers can change, but still electric charge is conserved. For example, when an electron (e^-) meets its antiparticle, called a positron (e^+),[3] they can annihilate, producing a pair of photons (γ): $e^- + e^+ \rightarrow 2\gamma$. In this example, both sides of the equation are also electrically neutral.

There is another quantity, called entropy, which is not conserved in nature, but has a profound impact on reactions. Entropy is a measure of statistical disorder; it quantifies the randomness of the molecules that make up a substance.

Two fundamental laws of thermodynamics relate to energy and entropy. The first law of thermodynamics expresses conservation of energy as it relates to heat flow. According to the second law of thermodynamics, the total entropy of the system plus the surroundings can either remain constant or increase, but it cannot decrease. The total energy of the system and surroundings does not change, while the total entropy is nondecreasing. One law expresses equality, the other inequality (greater than or equal to).

Many reactions are observed to occur only one way in nature – i.e. they are evidently irreversible. For example, heat spontaneously flows from high temperature to low temperature, but never spontaneously flows the opposite direction (by doing work, it can be "driven" to flow the other way – in contrast to occurring "spontaneously"). An ice cube melts on a hot summer day, whereas water does not freeze on a hot summer day. Both of these physical reactions satisfy the first law of thermodynamics, but only the melting of an ice cube on a hot summer day satisfies the second law of

---

[3] A positron is just like an electron with the same mass, except that all of its quantum numbers – like electric charge – are reversed (so the positron is positively charged). Electrons are abundant in nature – every atom has them (except for completely ionized atoms, like H^+) – but positrons are not. When positrons are produced, they don't last long because as soon as they come across any of the many electrons in matter, they tend to annihilate very quickly.

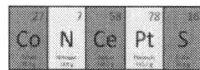

thermodynamics. The reverse reaction – a cup of water freezing on a hot summer day – does not occur in nature because it would lead to an overall decrease in the entropy – the statistical disorder of the molecules – of the system plus the surroundings.

An example of a chemical reaction that occurs spontaneously only one way is dissolving sugar in a cup of tea. When you drop a lump of sugar into a cup of tea and stir it, the sugar dissolves in the tea. Just try stirring the tea to try to reproduce the lump! This reverse reaction does not occur spontaneously because it would decrease the overall entropy (disorder) of the system and surroundings.

Chemical reactions are generally reversible – i.e. if $A + B \rightarrow C + D$ occurs, then $C + D \rightarrow A + B$ can be made to occur, in principle. If the entropy of the system plus surroundings increases one way, then the reverse reaction must be driven by doing work – i.e. it won't occur spontaneously. Energy is conserved both ways. If the reaction is exothermic one way, then the reverse reaction will be endothermic – and conservation of energy dictates that the energy released in the exothermic reaction must be exactly the same as the energy absorbed in the reverse endothermic reaction.

Another important property of a chemical reaction is the rate at which it occurs. There are many practical applications of chemistry where it is desirable to increase the rate of a chemical reaction. For example, it is

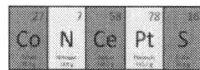

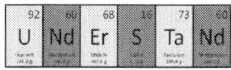 

generally desirable to speed up chemical reactions involved in healing. There are four common ways to speed up a reaction:

- Increasing the temperature increases the frequency of collisions, which increases the reaction rate. This is why it is more effective to wash your hands with warm water instead of cold water, for example.
- Increasing the concentration of the reactants also increases the frequency of collisions, and so this also increases the reaction rate. For example, burning occurs more rapidly in pure oxygen (O) than in air, since pure diatomic oxygen gas (O_2) has five times the concentration of oxygen as air (at the same pressure).
- Increasing the surface area of the reactants increases the reaction rate, as the reaction occurs at the surface where the reactants are in contact. This is why granulated sugar dissolves faster than a lump of sugar, for example.
- Adding a catalyst can increase the reaction rate. A catalyst is a substance that increases the rate of a reaction, but which does not undergo a chemical change during the reaction. In photosynthesis, chlorophyll (a green-colored pigment found in plants) serves as a catalyst, for example.

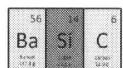 

Ch. 4 Self-Check Exercises

1. Determine how many atoms of each type there are in each of the following expressions:

Al_4C_3	$C_{12}H_{22}O_{11}$	$Mg(ClO_3)_2$
$4N_2O_3$	$3H_2SO_4$	$5C_2H_6$
$4Sn(NO_3)_2$	$2Ba(NO_3)_2$	$5Hg_3(PO_4)_2$

$Al(NO_3)_3$	$(NH_4)_2Cr_2O_7$
$8K_2CO_3$	$6H_3PO_4$
$4(CH_3)_2N_2H_2$	$3(NH_4)_2SO_3$

2. Balance the following chemical reactions:

$Sn + Cl_2 \rightarrow SnCl_4$	$H_2S + O_2 \rightarrow H_2O + SO_2$
$Br_2 + F_2 \rightarrow BrF_3$	$Al_2O_3 + HCl \rightarrow AlCl_3 + H_2O$
$Al + O_2 \rightarrow Al_2O_3$	$Na + Fe_2O_3 \rightarrow Fe + Na_2O$
$Rb + S_8 \rightarrow Rb_2S$	$Al + Fe_3O_4 \rightarrow Al_2O_3 + Fe$

$H_2S + HNO_3 \rightarrow S + NO + H_2O$
$NH_3 + O_2 \rightarrow NO + H_2O$
$CaO + P_4O_{10} \rightarrow Ca_3(PO_4)_2$
$Al_2(SO_4)_3 + Ca(OH)_2 \rightarrow Al(OH)_3 + CaSO_4$

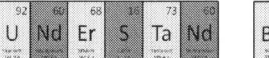 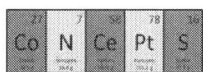

5 Phases of Matter

5.1 Phase Transitions

 e three basic phases of matter include solid, liquid, and gaseous. The term fluid is used to describe both liquids and gases. This is easy to remember if you associate fluids with substances that "flow."

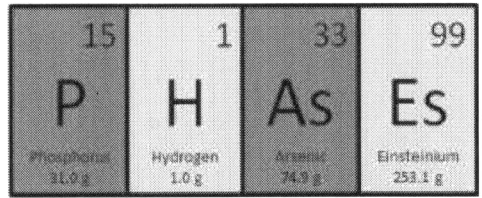

Note that when we go to the gas station, the "gas" that we buy is gasoline, which is liquid – not gaseous. When we use the word "gas" in this book, it will always be short for gaseous, and not for gasoline.

There is actually a fourth phase, called plasma. Whereas a gas consists of neutral particles, a plasma is composed of ions (charged particles). The sun is a plasma, for example. The temperature and pressure of the sun are so high that very frequent, energetic collisions cause atoms – mostly hydrogen (**H**) and helium (**He**) – to gain or lose electrons and exist as ions rather than neutral atoms.

The distinction between the three phases is physical – not chemical. It has to do with the strength of the intermolecular forces. The forces that hold the molecules of a solid are very strong, and so a solid tends to have a definite shape, whereas liquids and gases flow readily. The bonds are even weaker in gases, which flow more than liquids.

A substance may change from one phase to another, depending on its pressure and temperature. The following terms describe six possible phase transitions:

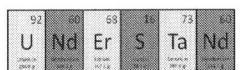 Freezing is a phase transition from liquid to solid.

Melting is a phase transition from solid to liquid.

Condensation is a phase transition from gas to liquid.

Vaporization is a phase transition from liquid to gas.

Deposition is a phase transition directly from gas to solid, skipping the liquid phase.

Sublimation is a phase transition directly from solid to gas, skipping the liquid phase.

Note that evaporation is <u>not</u> a phase transition. There is an important distinction between evaporation and vaporization. Vaporization is a change of phase from liquid to gas. When water (H_2O) boils, liquid water turns into steam. At standard pressure, water boils at 100°C. Boiling (vaporization) occurs throughout the entire volume of the liquid. Evaporation is not a change of phase. Evaporation occurs when the fastest molecules escape the liquid into the air. Evaporation occurs at all temperatures, not just the boiling point. Evaporation occurs only on the surface of the liquid.

A phase transition is a physical change, not a chemical change. A phase transition is brought about through changes in pressure and temperature, which cause changes in the strengths of the intermolecular forces. It is not a chemical change because it does not change the identity of the molecules. For example, at standard pressure, liquid water (H_2O) freezes into ice at 0°C and becomes steam when it boils at 100°C. In contrast, when a cup of water evaporates, no phase transition occurs – this results in water vapor in the air, rather than steam.

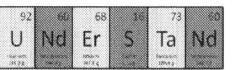 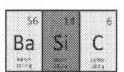

Heat is either absorbed or released by a substance that undergoes a phase change. Heat must be added to a solid in order to change it into a liquid (melting) or gas (sublimation), and must be added to a liquid in order to change it into a gas (vaporization). Melting, sublimation, and vaporization are endothermic physical reactions. Heat is instead released by a gas that changes into a liquid (condensation) or solid (deposition), and is released when a liquid changes into a solid (freezing). Condensation, deposition, and freezing are exothermic physical reactions.

5.2 Properties of Matter

 e will discuss various properties of matter – solid, liquid, or gaseous – that can be measured in this section. Many of the quantities that we consider here will be useful in subsequent sections of this chapter.

One quantity that you can measure is temperature. The temperature of a substance provides a measure of the average kinetic energy of its molecules. Kinetic energy equals $\frac{mv^2}{2}$, where m is mass and v is speed. The faster the molecules move, on average, the greater the temperature of the substance.

The SI unit of temperature is the Kelvin (K). Other temperature units – Celsius (°C) and Fahrenheit (°F) – are not SI units. Some formulas only work if the temperature is expressed in Kelvin. The temperature in Kelvin is called

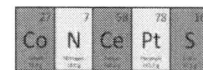

the absolute temperature. We use the degree symbol (°) for Celsius and Fahrenheit, but not for Kelvin.

Use the following formulas to convert between Celsius, Fahrenheit, and Kelvin:

$$T_K = T_C + 273.15$$
$$T_C = T_K - 273.15$$
$$T_F = \frac{9}{5}T_C + 32$$
$$T_C = \frac{5}{9}(T_F - 32)$$

Convert 30°C to Fahrenheit.

$$T_F = \frac{9}{5}T_C + 32 = \frac{(9)(30)}{5} + 32 = 54 + 32 = 86°F$$

Convert 50°F to Celsius.

$$T_C = \frac{5}{9}(T_F - 32) = \frac{5}{9}(50 - 32) = \frac{5}{9}18 = 10°C$$

Convert 70°C to Kelvin.

$$T_K = T_C + 273.15 = 70 + 273.15 = 343\ K$$

Convert 300 K to Celsius.

$$T_C = T_K - 273.15 = 300 - 273.15 = 27°C$$

If you want to convert Kelvin to Fahrenheit, first convert Kelvin to Celsius and then convert Celsius to Fahrenheit. If you want to convert

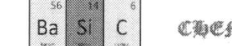 
Fahrenheit to Kelvin, first convert Fahrenheit to Celsius and then convert Celsius to Kelvin.

A couple of special temperatures to know are the freezing and boiling points of water (H_2O). At standard pressure, water freezes at 0°C and boils at 100°C. In Kelvin, these temperatures are 273.15 K and 373.15 K, and in Fahrenheit they are 32°F and 212°F. In the United States, weather forecasters often state the temperature in Fahrenheit.

Room temperature or standard temperature are not exactly standard – since there are a few different common conventions – but generally refer to a temperature of 20°C to 25°C – which is 293.15 K to 298.15 K or 68°F to 77°F.

Heat is a transfer of thermal energy. The direction of the heat flow tends to be from higher temperature to lower temperature. For example, if you touch an ice cube, heat is transferred from your higher-temperature finger to the lower-temperature ice cube.

Pressure (P) is defined as force (F) per unit area (A): $P = F/A$. Pressure considers not only the force being exerted, but also how the force is applied. When the same force is spread out over a larger area, there is less pressure, and when it is concentrated over a smaller area, there is more pressure. It hurts to receive a shot because although the force is very small, the area of the needle's head is very tiny.

The SI unit of pressure is the Pascal (Pa), which equates to one Newton per square meter (N/m^2). Some other common units include atmospheres (atm), pounds per square inch (psi), millimeters of mercury (mm **Hg**), millibars (mbar), and the torr (named for Toricelli).

A 200-lb. football player with flat-soled shoes and a 100-lb. cheerleader with high heels are dancing. Who is more likely to damage the dance floor?

Pressure is force per unit area: $P = \dfrac{F}{A}$. The football player exerts twice as much force on the floor, but the cheerleader's high heel has much less area. The **cheerleader**'s shoe exerts greater pressure on the floor, and is more likely to damage the floor.

How can it be comfortable to lie down on a bed of nails? Why might it be painful to stand on a bed of nails?

Each nail has very little area. Standing on a single nail would be very painful. But lying down on a bed of nails, the same force (weight) is distributed over hundreds of nails, which greatly increases the area, thereby reducing the **pressure**. Standing on a bed of nails, however, involves less area, and more pressure.

A scale measures weight (W), which is the force of gravity that the earth exerts on an object. The SI unit of weight, like all forces, is the Newton (N), which equates to one $\text{kg} \cdot \text{m/s}^2$. Weight can <u>not</u> be expressed in grams (g) or kilograms (kg) – those are units of mass (m). However, mass and weight are related through the formula $W = mg$, where g is gravitational acceleration – which equals 9.81 m/s^2 near the surface of the earth. Many scales which directly measure weight give mass readings in grams. In this case, the manufacturer of the scale has calibrated the scale for use near the surface of the earth by effectively dividing the weight by 9.81 m/s^2 in order to provide the mass (since $m = W/g$). If you take the same scale to the moon (where $g = 1.6 \text{ m/s}^2$), the measurements will be incorrect. The mass of an object is the same regardless of its location, whereas its weight depends on the strength of gravity. For example, a rock would have the same mass but about one-sixth of its weight if it were transported from the earth to the moon.

There are a couple of different types of mass in chemistry. There is the total mass of the object, which we will denote by a lowercase m, not to be confused with the molar mass, which we will denote by an uppercase M.[4] The molar mass, M, is not the total mass of the object, but the mass of 1 mole (see Sec. 1.4) of the substance (often expressed in g/mol). The molar mass is usually what you find on a periodic table.

Density (d) is defined as mass per unit volume (V): $d = m/V$. Volume refers to the amount of space than an object takes up. Density is a property of a material, and does not depend on how much of the material you have. For example, if you cut a wooden block in half, each half of the block has one-half of the original mass and volume, but the same density as the original block. If you compare a chunk of steel to a chunk of wood, you will notice that steel is more dense than wood. Steel is not necessarily heavier than wood – if you have a small chunk of steel and a very large chunk of wood, the chunk of wood may be heavier. However, if you have a chunk of

---

[4] Unfortunately, not all books or instructors use the same notation, so if you are taking a course, you might use different symbols.

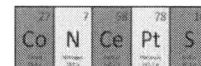
wood and a chunk of steel of the same size, then the chunk of steel will be heavier because steel is more dense than wood.

5.3 Ideal Gases

I types of fairly low-density gases exert similar physical behavior, with little to no regard for their chemical composition. We call such low-density gases "ideal gases," and they are the simplest to model quantitatively. The behavior of higher-density gases is somewhat more complicated. In this book, we will restrict our attention to ideal gases.

Ideal gases satisfy the following laws. According to Boyle's law, the pressure (P) and volume (V) of an ideal gas are inversely proportional at constant temperature:

$$P_1 V_1 = P_2 V_2$$

Boyle's law says that an increase in pressure tends to decrease the volume of an ideal gas if its temperature is constant. According to Charles's law, the temperature (T) and volume of a gas are directly proportional at constant pressure:

$$\frac{V_1}{T_1} = \frac{V_2}{T_2}$$

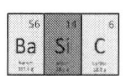 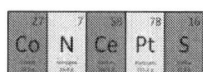
Charles's law says that an increase in temperature tends to increase the volume of an ideal gas if its pressure is constant. Boyle's and Charles's laws are special cases of the ideal gas law:

$$\frac{P_1 V_1}{T_1} = \frac{P_2 V_2}{T_2}$$

In these equations, the temperature must be expressed in Kelvin (K). If a problem gives you the temperature in Celsius or Fahrenheit instead, first convert the temperature to Kelvin before plugging it into any of these formulas.

The ideal gas law can alternatively be expressed as $PV = NRT$, where N is the number of moles and R is the universal gas constant. In SI units, $R = 8.31 \frac{J}{mol \cdot K}$, but SI units are not convenient in typical chemistry calculations. Thus, most chemistry students know it as $R = 0.0821 \frac{L \cdot atm}{mol \cdot K}$, since pressure is often measured in atmospheres (atm) instead of Pascals (Pa) and volume is often measured in liters (L) instead of cubic meters (m^3).

The atmosphere is a very large gas with which we are all familiar. The composition of earth's atmosphere is predominantly diatomic nitrogen (N_2), at 78%, and diatomic oxygen (O_2), at 21%. The remaining 1% includes argon (Ar), carbon dioxide (CO_2), and other gases.

The initial volume of a gas is 20 cubic centimeters (cc). The pressure of the gas doubles at constant temperature. What is the final volume?

According to Boyle's law, at constant temperature $p_1V_1 = p_2V_2$. If the pressure doubles, the volumes halves. That is, if $p_2 = 2p_1$, then $p_1V_1 = 2p_1V_2$ so that $V_2 = V_1/2$. The final volume is **10 cc**.

The initial temperature of a gas is 27°C. The volume of the gas doubles at constant pressure? What is the final temperature?

According to Charles's law, at constant pressure $\frac{V_1}{T_1} = \frac{V_2}{T_2}$.

If the volume doubles, the temperature also doubles – but only when the temperature is expressed in Kelvin. The final temperature is **600 K** (since the initial temperature was 300 K).

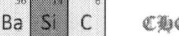 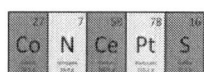
5.4 Classes of Solids

ere are two types of solids – amorphous and crystalline. The distinction has to do with whether or not the atoms are arranged in a geometric pattern.

The atoms of amorphous solids do not come in regular arrangements. The atoms of crystalline solids appear in repeated geometric patterns.

Examples of amorphous solids include glass and plastic. Amorphous solids soften gradually when heated and so do not have a specific melting point.

In crystals, the melting point occurs when the bonds are suddenly broken by an increase in the thermal energy (heat). There are four types of crystalline solids:

- The bonds of ionic crystalline solids involve a transfer of electrons. For example, salts like sodium chloride (**NaCl**) are ionic crystals. In this bond, a metal transfers electrons to a nonmetal, and the oppositely charged ions (**Na$^+$** and **Cl$^-$**, in the case of **NaCl**) attract one another.

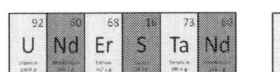 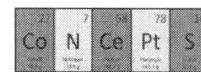
Ionic crystals are hard and have high melting points because the bonds are quite strong.

- The bonds of covalent crystalline solids involve sharing electrons. Examples of covalent crystals include diamond and graphite. These are two different forms of carbon (**C**), which bonds covalently: Carbon atoms naturally share electrons with one another, as carbon has 4 valence electrons. Diamond is hard and shiny; graphite is soft and black, also with a luster. Covalent crystals are hard and have a very high melting point as the bonds – like ionic crystals – tend to be quite strong.

- Metallic crystals have a sea of electrons that can flow easily. Examples include metals like aluminum (**Al**) and copper (**Cu**). Metallic crystals are good conductors of heat and electricity, are malleable, and have a luster.

- Molecular crystals use van der Waals forces in order bind stable molecules together. An example of a molecular crystal is ice (which consists of stable H_2O molecules). Compared to other crystalline solids, molecular crystals tend to be soft and have a low melting point.

5.5 Liquids

ressure (P) in a fluid (liquid or gas) varies with depth (h) according to the equation $P = dgh$, where d represents the density of the fluid and g is gravitational acceleration – which equals 9.81 m/s^2 near earth's surface. In the formula $P = dgh$, it is important to realize that the symbol h represents depth, measured from the top of the fluid, as illustrated on the following page. (It is _not_ height measured upward from the bottom.)

 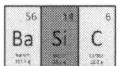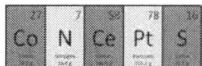

Since pressure in a fluid increases with greater depth, when an object is submerged in a fluid, there is greater pressure at its bottom and less pressure at its top. The result is that the fluid exerts a net upward force on the object. We call this a buoyant force.

Measuring depth

In the formula $P = dgh$, the symbol h represents **depth**, measured from the top of the fluid, as illustrated below. (It is **not** height measured upward from the bottom.)

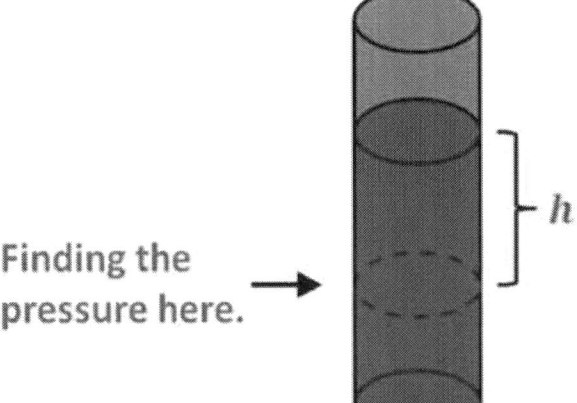

Finding the pressure here.

h

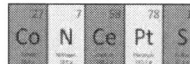
In which container illustrated below is there greatest pressure at the bottom of the red liquid?

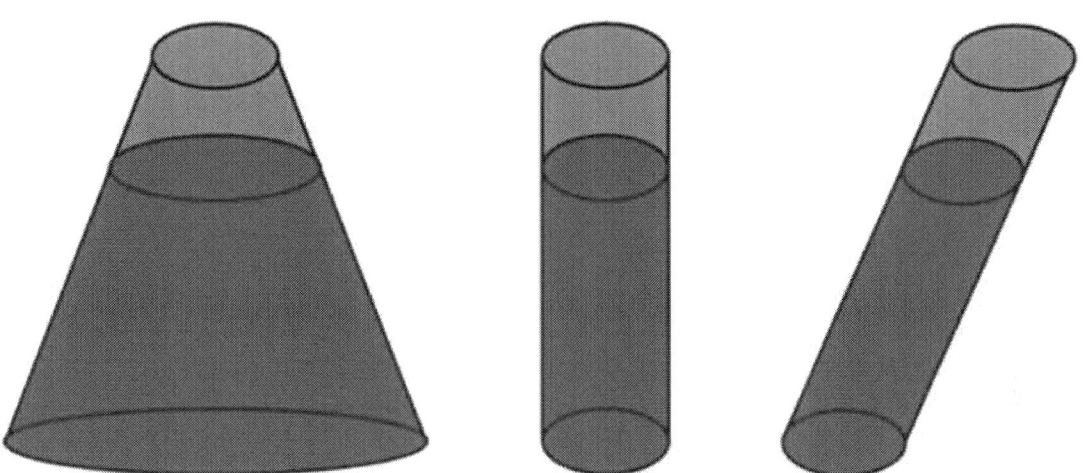

The depth below the top of the liquid is the same for each. All three contain the same liquid, so the density is also the same for each. Therefore, all three containers have the same **pressure** at the bottom, since $P = dgh$.

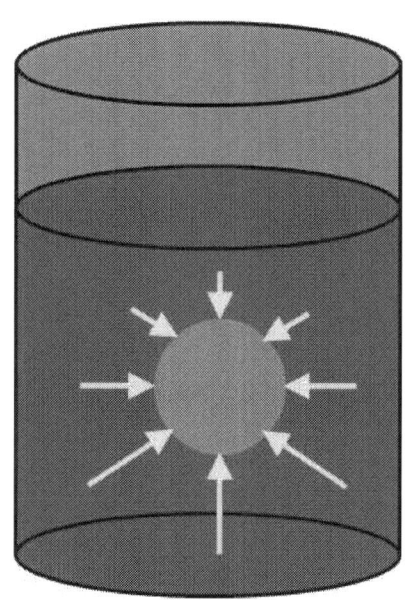

 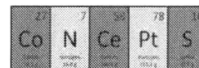

According to Archimedes' principle, the upward buoyant force exerted on a submerged object is equal to the weight of fluid that it displaces. The buoyant force equals the weight of the displaced fluid, which is generally <u>not</u> equal to the weight of the object.

Weight (the force of gravity) pulls downward on a submerged object, while the buoyant force (from Archimedes' principle) pushes upward on the submerged object. The buoyant force only equals the weight of the object if the object is floating. If the object is rising or sinking, the buoyant force is greater than or less than the object's weight, respectively. No matter what, the buoyant force always equals the weight of the displaced fluid.

If an object is more dense than the fluid, the object will sink (like a block of lead falling in water). If an object is less dense than the fluid, the object will rise (like an ice cube rising upward in a cup of water). If an object is just as dense as a fluid, the object will float – i.e. neither sink nor rise. An object that is less dense than a fluid will also float once it rises to the top.

A helium balloon rises upward because helium is less dense than air. A boat made of steel could float on water even though steel is much more dense than water – because all of the air inside the boat can make the overall density of the boat less dense than water. In this case, the boat must have a lot of volume of air in order to lower its overall density.

Water (H_2O), by far the most abundant liquid on earth (covering about 70% of earth's surface), has a density of approximately 1 g/cc (0.998 g/cc) at STP. It is useful to compare the density of other objects to water. As such, there is a quantity called specific gravity, which is defined as the density of a substance divided by the density of water. The specific gravity of water is therefore exactly 1.

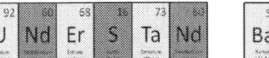 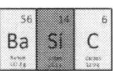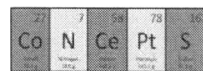
For example, lead (**Pb**) has a specific gravity of about 11. Lead is more dense than water since its specific gravity is greater than 1; lead sinks in water. Oil, on the other hand, has a specific gravity that is less than 1 because oil is less dense than water; that's why oil floats on water.

When water is stored in a glass container – such as a beaker or a graduated cylinder – it is curved at the top: The outer edge is higher than the center. This is called a meniscus. This concave meniscus forms because of surface tension – the water molecules attract to the molecules of the glass container.

Water exhibits strange physical behavior as it freezes. Most materials expand when heated. For example, a metal rod becomes slightly longer when it is heated. Water is one of the few substances that actually expands when it is cooled from liquid water to ice. This is why pipes sometimes burst on a very cold winter night. When the water freezes into ice inside the pipes, the water expands. This also explains why ice cubes float on water. The ice cubes occupy more volume than they did as water, and therefore have less density than water. For most substances, the solid form is much more dense than the liquid form. Water is an exception to the rule.

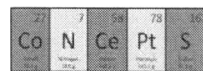

 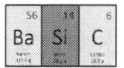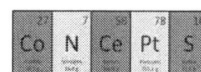
5.6 Solutions

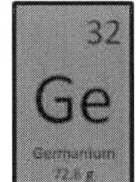 nerally, a solution is any homogeneous mixture (uniform throughout) of two or more substances. Thus, a solution could be a gas (like the air), a liquid (like a cup of tea), or a solid (like brass, which is an allow made of zinc, **Zn**, and copper, **Cu**). In this section, we will focus exclusively on liquid solutions.

A solution consists of a solute (pronounced "sol-yoot") and a solvent. The solute is the substance dissolved in the solution, while the solvent is the substance that dissolves the solute. When a solid or gas is dissolved in a liquid, the liquid is the solvent and the solid or gas is the solute. Water (H_2O) often serves as the solvent because it is very abundant on earth and also because it is highly effective as a solvent.

The concentration of a solution equals the amount of solute divided by the amount of solvent. For a given amount of solvent, more solute means a greater concentration. There are two common, yet different, ways of expressing the concentration of a solution – molarity and molality. Since the two terms are almost spelled the same, it's very important to pay attention to the distinction. The molarity of a solution equals the number of moles of the solute per liter of solution, whereas the molality of a solution equals the number of moles of solute per kilogram of solution. One involves capacity (volume), while the other involves mass.

The solubility of a solution equals the maximum concentration of solute that can be dissolved in a solution; any excess solute will remain undissolved. A solution that has reached its solubility is said to be saturated. If instead

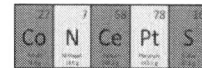

additional solute can be added and dissolved in a solution, the solution is said to be unsaturated.

It is actually possible for a solution to become supersaturated. This is possible because the solubility of a solution increases with temperature. That is, more solute can be dissolved in a solution at higher temperature than at lower temperature. By slowly cooling a saturated solution, the solution can become supersaturated – containing a greater concentration of solute than would normally occur at that temperature. As an example, you can dissolve more sugar in hot tea than iced tea. By slowly cooling hot tea, you can create a supersaturated solution of iced tea – i.e. you can make iced tea with more sugar dissolved than would otherwise be possible.

Some liquids are polar, while others are nonpolar. The molecules of a polar liquid behave like dipoles: Each neutral molecule can be split in half such that there is a net positive charge on one side and a net negative charge on the other side. Water (H_2O), for example, is a polar liquid: Two hydrogen (H) atoms bond covalently with one oxygen (O) atom in a single water molecule, where the O atom shares an electron with each H atom. In the figure on the next page, there is a net negative charge below the dashed line because each H atom has 1 proton and there is a net positive charge above the dashed line (only the valence electrons are shown – the O atom also has 2 more electrons in the inner shell, which are not shown in the figure). The charge distribution of the H_2O molecule looks like the dipole illustrated on the right side of the figure on the following page.

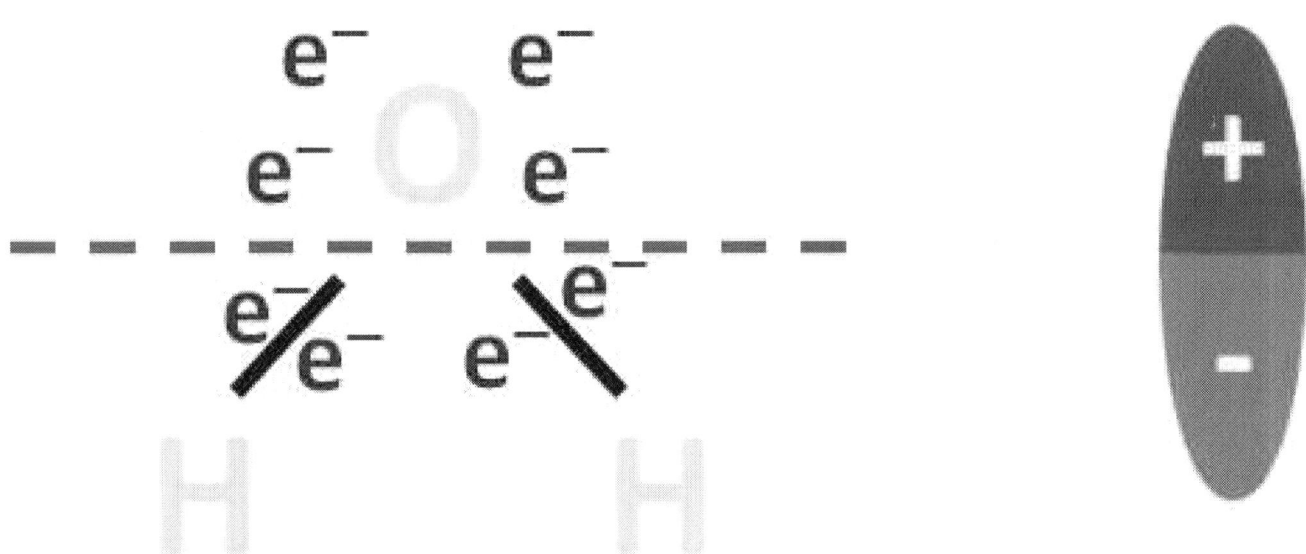

In a nonpolar liquid, the charge is evenly distributed in each molecule. Gasoline is an example of a nonpolar liquid.

Whether or not the molecules of a liquid are polar or nonpolar plays a very significant role in the behavior of the liquid in solutions. In particular, it is generally the case that "like dissolves like," where "like" refers to the bonding of the molecules of the solute and solvent. We will explain what this means in the next two paragraphs.

Polar liquids, like water, consist of clumps of molecules because polar molecules attract one another electrically – i.e. the negative part of one polar molecule attracts to the positive part of another polar molecule. Salts, like sodium chloride ($NaCl$), have ionic bonds, which are also dipoles – they consist of positive ions attracted to negative ions (like Na^+ and Cl^- in $NaCl$). Like water, the molecules of salts tend to clump together. Water molecules interact with salt molecules because charged parts of water molecules interact with the charged ions of salts. In this way, water is able to dissolve salts. The molecules of sugars clump together like salts and water, and so sugar also dissolves readily in water. Salts and sugars readily dissolve in polar liquids like water because they involve similar bonding (i.e. the clumping together of molecules due to opposite charges).

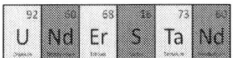 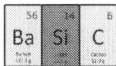 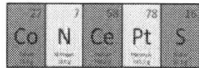

The molecules of nonpolar liquids, like gas and oil, do not tend to clump together like the molecules of polar liquids because their charges are evenly distributed. The molecules of fats are nonpolar. Fats do not dissolve well in polar liquids like water, but do dissolve readily in nonpolar liquids like gasoline. This is what we mean by "like dissolves like."

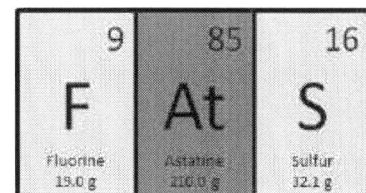

Some compounds, called electrolytes, dissociate – i.e. they separate into positive and negative ions – when they are dissolved in water. For example, salts – like sodium chloride ($NaCl$) – are electrolytes. When $NaCl$ dissolves in water, it dissociates into Na^+ and Cl^-: $NaCl \rightarrow Na^+ + Cl^-$. Solutions with electrolytes can conduct electric current. This property of electrolytic solutions is useful for electrochemistry, as described briefly toward the end of Sec. 4.3.

Substances that do not dissociate in solutions are called nonelectrolytes. Sugar is an example of a nonelectrolyte – it does not separate into positive and negative ions when it dissolves in water.

5.7 Acids and Bases

ids are often ionic compounds that contain hydrogen (**H**) and which dissociate into H^+ along with a negative ion in an aqueous solution – i.e. a solution that contains water. For example, hydrochloric acid (**HCl**) dissociates into H^+ and Cl^- in a solution of water. Most foods that taste sour, including lemons and vinegar, get their sour taste from an acid.

Hydrogen is more weakly bonded in strong acids, and more strongly bonded in weak acids. Thus, strong acids dissociate more completely than weak acids (it is easier to break the weak bonds of strong acids). Examples of strong acids include sulfuric acid (H_2SO_4), which is found in batteries, and hydrochloric acid (HCl), which is a gastric acid. Examples of weak acids include coffee and tomatoes.

Many common bases are ionic compounds that contain hydroxide (OH^-) and dissociate into OH^- along with a positive metal ion in an aqueous solution. For example, sodium hydroxide (NaOH) dissociates into Na^+ and OH^- in a solution of water. Whereas acidic foods taste sour, bases that are not too harmful to taste, like soap, taste bitter.

Hydroxide is more weakly bonded in strong bases, and more strongly bonded in weak bases. Thus, strong bases dissociate more completely than weak bases (it is easier to break the weak bonds of strong bases). Examples of strong bases include many household cleaners, such as oven cleaner, which has sodium hydroxide (NaOH). Examples of weak bases include baking soda and soap. (Note that soda that you drink – i.e. cola – is acidic, whereas baking soda is basic.)

The pH scale tells you if a solution is a strong acid, weak acid, neutral, weak base, or strong base. The letters pH are an abbreviation for the French phrase for "hydrogen power," which is "pouvoir hydrogène." The pH of a solution is a number between 0 and 14. A solution with a very low pH is a very strong acid, like battery acid and gastric fluid. A solution with a pH just below 7 is a weak acid, like tomatoes and coffee. Water is neutral, with a pH of 7. A solution with a pH just above 7 is a weak base, like soap and baking soda. A solution with a very high pH is a strong base, like oven cleaner.

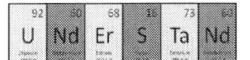   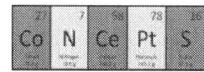

Ch. 5 Self-Check Exercises

1. For each of the following pairs of substances, indcate which is more dense: aluminum/wood, air/ice, lead/wood.

2. Are you more likely to pop a balloon by punching it hard with your fist or gently pressing a needle into it. Which concept explains this?

3. A 100-m diameter pond has a 50-m depth. A 50-m diameter lake has a 100-m depth. Which has greater pressure at the bottom?

4. Blocks of wood, aluminum, and lead have the same volume. Which weighs more?

5. Blocks of wood, aluminum, and lead have the same mass. Which is largest?

6. One aluminum rod is twice as long as another. Which is more dense?

7. Convert each of the following to Fahrenheit: 80°C, 200 K.

8. Convert each of the following to Celsius: 120°F, 400 K.

9. Convert each of the following to Kelvin: 10°C, 20°F.

10. The initial volume of an ideal gas is 20 cc. The pressure of the gas halves at constant temperature. What is the final volume of the gas?

11. The initial temperature of an ideal gas is 23°C. The volume of the gas halves at constant pressure. What is the final temperature of the gas?

12. Indicate whether each of the following solids is amorphous or a covalent, ionic, metallic, or molecular crystal: aluminum, diamond, sodium chloride, rubber, dry ice (CO_2).

13. When you add sugar to tea, which substance is the solute and which substance is the solvent?

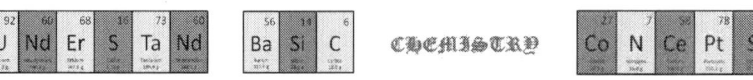

14. Indicate whether each of the following is an acid or a base: lemon juice, soda, oven cleaner, H_2SO_4, $Ba(OH)_2$.

15. Indicate whether each of the following solutions is a strong acid, weak acid, neutral, weak base, or strong base: pH of 9, pH of 2, pH of 7, pH of 13, pH of 5.

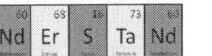 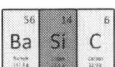
6 At Atoms and Nuclei

6.1 Atomic Structure

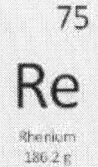

call that atoms consist of protons, neutrons, and electrons. Protons and neutrons reside in a tiny space in the nucleus of an atom: The size of a typical atomic nucleus is about one femtometer (fm), which is 10^{-15} m. That's 0.000000000000001 m, or a trillionth of a millimeter. The electrons reside in a much larger – yet still very tiny – space surrounding the nucleus, in probability clouds called shells and subshells. The size of a typical atom is about one Angstrom (Å), which is 10^{-10} m. That's 0.0000000001 m, or less than a millionth of a millimeter.

The overall size of an atom is about a hundred thousand (100,000) times larger than the nucleus. If you made a model of an atom the size of a football field, the nucleus would be about the size of a pea. Since electrons are "pointlike" particles – i.e. they are elementary particles, evidently not composed of other smaller particles – the vast majority of an atom is really just empty space. Think about this: Matter is mostly made of empty space! However, a person made of mostly empty space can't walk through a wall made of mostly empty space because the bonds between the atoms of walls and people are very strong.

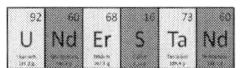 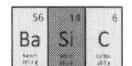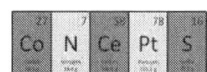
How large is a typical atom?

A typical **atom** is about the size of one Angstrom.

- An Angstrom is 10^{-10} m.

10^{-3} = milli
10^{-6} = micro
10^{-9} = nano

How large is a typical nucleus?

A typical **nucleus** is about the size of a femtometer.

- A femtometer is 10^{-15} m.

An atom is 100,000 times larger than its nucleus!

Matter is mostly made of empty space!

Electrons are elementary particles. They are pointlike.

But I can't walk through walls!

The nucleus is positively charged – as protons are positive and neutrons are neutral – which attracts the electrons – which are negatively charged – that surround the nucleus. The protons and neutrons have nearly the same mass – neutrons are slightly heavier – and each have much more mass than the electrons.

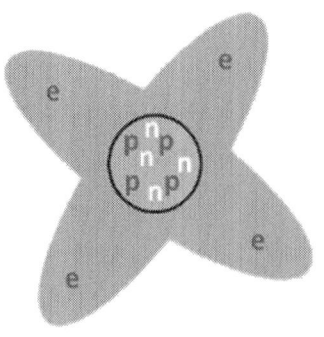

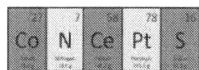
The number of protons determines the identity of the atom. For example, an atom that has 4 protons in the nucleus is definitely beryllium (**Be**). The number of protons – also called the atomic number – is listed next to each element on the periodic table: Hydrogen (**H**) has 1 proton, helium (**He**) has 2 protons, lithium (**Li**) has 3 protons, and so on.

If an atom is neutral, the number of electrons is the same as the number of protons. If instead the atom is an ion – i.e. it has a net electric charge – then it has more or fewer electrons than protons. For example, Li^+ is a lithium (**Li**) ion with 3 protons, but only 2 electrons, and O^{2-} is an oxygen (**O**) ion that has 8 protons, but 10 electrons. (We are counting the total number of electrons in these examples, not just the valence electrons.)

The electrons reside in probability clouds called orbitals, which are arranged in shells and subshells. The electrons do not travel in orbits that are well-defined circles (as often illustrated in simple models), but orbit the nucleus in an orbital. The orbital is a region of space, called a probability cloud, that represents the possible locations of the electron. The motion of the electron is governed by a field of physics called quantum mechanics, which says that the behavior of electrons is subject to probability. One example of the probabilistic nature of electrons is Heisenberg's uncertainty principle, which states that it is impossible to know both the exact location and the exact momentum (mass times velocity) of the electron (or any other particle) at the same instant. The best that we can do is calculate (or measure) probability clouds (orbitals) that tell us where the electrons are allowed to reside. We can't find the exact position of the electron, but we can narrow it down to a bubble.

Each electron in an atom occupies a specific quantum state. The quantum state of an electron consists of four quantum numbers:

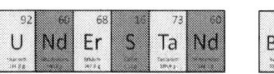 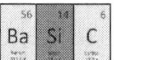 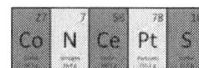
The principal quantum number, n, corresponds to the electron's main energy level. The value of n is an integer – 1, 2, 3, etc. – that determines the shells.

The orbital quantum number, ℓ, represents the magnitude of the electron's angular momentum – it indicates the shape of the orbital. The value of ℓ is an integer between 0 and $n - 1$ that determines the subshells. For example, for the energy level $n = 3$, the possible values of ℓ are 0, 1, and 2 (since $3 - 1 = 2$).

The magnetic quantum number, m_ℓ, specifies the direction of the electron's angular momentum – it indicates the orientation of the orbital. The value of m_ℓ is an integer between $-\ell$ and ℓ. For example, for $\ell = 2$, the possible values of m_ℓ are $-2, -1, 0, 1$, and 2.

The spin quantum number, m_s, relates to the direction of the electron's spin. Just as the earth has two types of angular momentum – i.e. it revolves around the sun in its annual orbit and also spins on its daily axis – electrons have spin angular momentum in addition to orbital angular momentum. The value of m_s can be $+\frac{1}{2}$ or $-\frac{1}{2}$, corresponding to the two ways that an electron can "spin."

Each value of the principal quantum number, n, designates a shell of the atom, which corresponds to a main energy level of the electrons. There are n^2 orbitals in each shell. For example, there is $1^2 = 1$ orbital in the first shell, there are $2^2 = 4$ orbitals in the second shell, there are $3^2 = 9$ orbitals in the third shell, and so on.

Each value of the orbital quantum number, ℓ, designates a subshell within the shell. There are n subshells within each shell. For example, there is just 1 subshell for $n = 1$, which we call the s sublevel; there are 2 subshells for $n = 2$, which we call the s and p sublevels; and there are 3 subshells for $n = 3$, which we call the s, p, and d sublevels.

The magnetic quantum number, m_ℓ, determines how many orbitals there are in a subshell. There are $2\ell + 1$ orbitals in a subshell, since m_ℓ can range from $-\ell$ to ℓ. For example, there is only 1 orbital in the s subshell ($\ell = 0$), there are 3 orbitals in the p subshell ($\ell = 1$), and there are 5 orbitals in the d subshell ($\ell = 2$).

According to the Pauli exclusion principle, no two electrons in an atom can occupy the same quantum state. The quantum numbers n, ℓ, and m_ℓ specify each orbital. There can be two electrons in a single orbital – one with spin up and one with spin down (the two possible values of m_s).

The first shell ($n = 1$) of an atom has just 1 subshell ($\ell = 0$) – the s sublevel – and holds up to 2 electrons (one with spin up and one with spin down). This is why the first period (row) of the periodic table has just 2 elements.

The second shell ($n = 2$) of an atom has 2 subshells ($\ell = 0$ and $\ell = 1$) – the s and p sublevels – and holds up to 8 electrons (since there is 1 s-orbital and there are 3 p-orbitals, and each orbital can hold 2 electrons – one spin up and one spin down). This is why the second and third periods of the periodic table have 8 elements.

The third shell ($n = 3$) of an atom has 3 subshells ($\ell = 0$, $\ell = 1$, and $\ell = 2$) – the s, p, and d sublevels – and holds up to 18 electrons (since there is 1 s-orbital, there are 3 p-orbitals, and there are 5 d-orbitals, and each orbital can hold 2 electrons – one spin up and one spin down). This is why the fourth and fifth periods of the periodic table have 18 elements.

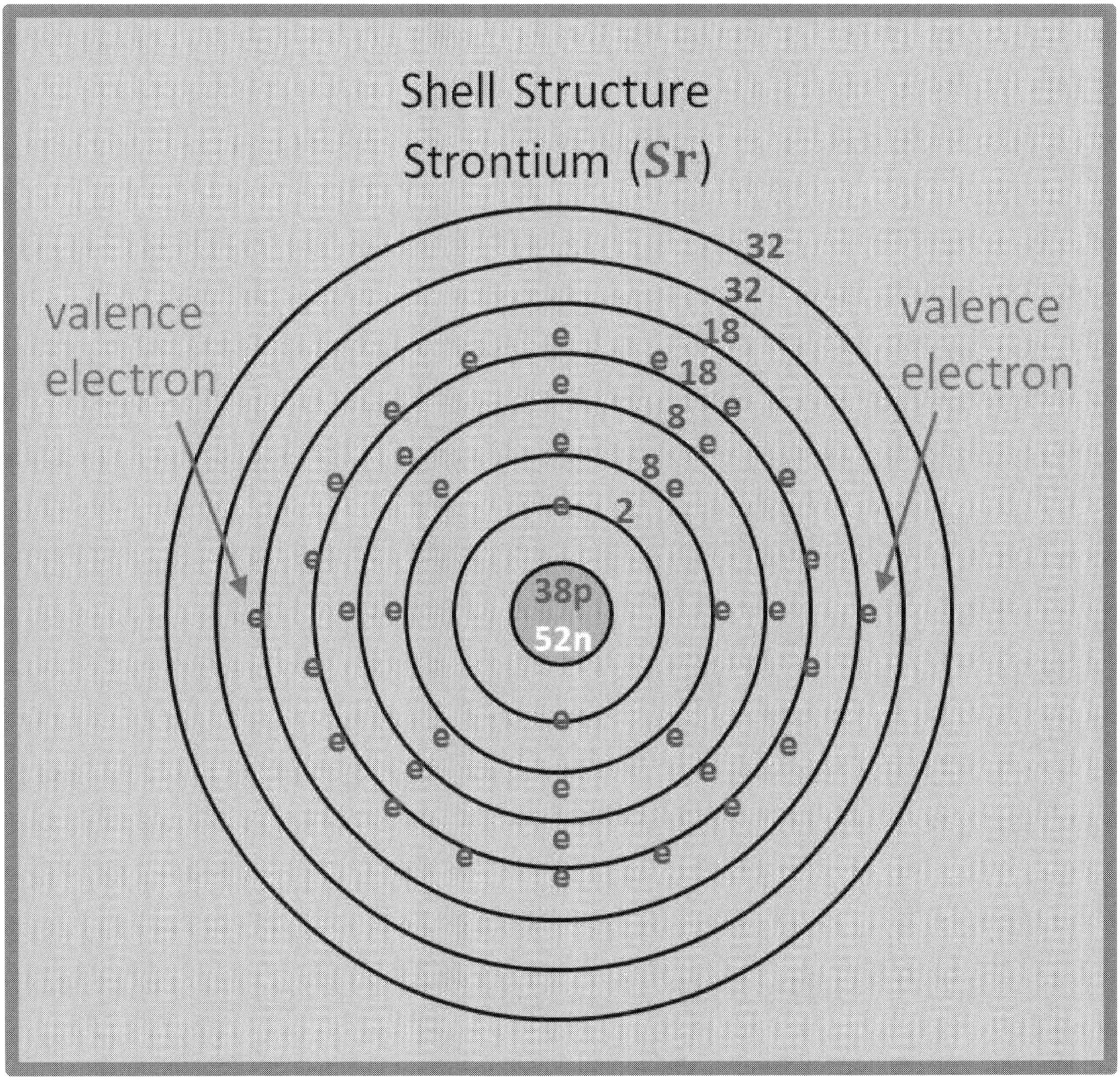

Shell Structure
Strontium (Sr)

valence electron

valence electron

32
32
18
18
8
8
2

38p
52n

The fourth shell ($n = 4$) has 4 subshells, which includes 7 f orbitals. Thus, there are 32 elements in the sixth and seventh periods of the periodic table (which include the lanthanides and actinides that appear below the table).

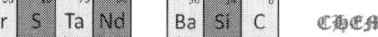 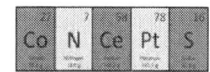

6.2 Nuclear Structure

ink about this: The electrons are attracted to the nucleus because electrons are negatively charged and the protons in the nucleus are positively charged – opposites attract. But what possesses those positively charged protons to reside together with electrically neutral neutrons in a very tiny nucleus – just 10^{-15} m in size? Like charges repel, meaning that protons repel one another electrically. If protons exert repulsive forces on one another electrically – i.e. they push each other away electrically – what holds them together in such a tiny nucleus?

The answer is the strong nuclear force. Protons attract other protons, and also neutrons, according to the strong nuclear force, while protons repel other protons according to the electric force. The electric force and strong nuclear force are two different forces that push or pull on the protons in the nucleus – the electric force causes protons to repel one another, while the strong nuclear force attracts both protons and neutrons together. Inside a nucleus, the attraction from the strong nuclear force is greater than the repulsion from the electric force, and so the strong nuclear force holds the nucleus together even though protons repel electrically.

There are four fundamental forces in nature: the gravitational force, electromagnetic force, strong nuclear force, and weak nuclear force. Any two objects with mass attract gravitationally. The force of gravity is the weakest force, so the gravitational attraction between two protons is insignificant. It takes an astronomical amount of mass – like the earth – to produce a significant force of gravity. Gravity explains the orbits of the planets around the sun and the motion of projectiles near earth, for example. Any two objects with electric charge attract or repel one another electromagnetically: like charges repel, and opposites attract. The strong nuclear force binds protons and neutrons together in the nucleus, and the weak nuclear force is responsible for radioactive decays – it allows for carbon-dating, for example.

Are electrons attracted to or repelled by protons?

Opposite charges attract.		
• This explains why the electrons are attracted to the nucleus.	− e	+ p

Are protons attracted to or repelled by one another?

Like charges repel one another.		
• The protons don't like each other electrically.	+ p	+ p

Then what binds protons together in the nucleus?

Why do protons reside in a tiny space if they don't like each other?	The strong nuclear force.

Gravity and the electromagnetic force follow inverse-square laws. These forces have infinite range, and get smaller as the separation between the objects increases. The nuclear forces, in contrast, are short-range: It is really strong if the separation is 10^{-15} m or less, but insignificant if the separation is greater. At a separation less than 10^{-15} m, the strong nuclear force is greater than the electric force. Within the size of a nucleus, the strong nuclear force attracts protons and neutrons together with a greater force than the electrical repulsion between protons.

An alpha particle is a helium nucleus — it consists of two protons and two neutrons. Do the protons of an alpha particle attract one another or repel one another?

They **attract** one another. Inside the nucleus, the protons are close enough that the strong nuclear force dominates, so they attract.

Do two different alpha particles attract one another or repel one another?

They **repel** one another. These protons are further apart, so their electrical repulsion dominates the strong nuclear force.

The atoms at the top of the periodic table have about as many protons as neutrons. The atoms at the bottom of the periodic table have many more neutrons than protons. Why?

Consider the question posed above on the right. If you look at the periodic table (see pages 12-13), the atomic number indicates the number of protons, while the approximate number of neutrons can be deduced (we will learn more about this in the next section) by subtracting the atomic number from the molar mass because most of the mass of an atom comes from the protons and neutrons (electrons are very light) and since protons and neutrons have nearly the same mass.

For example, a helium (**He**) atom typically has 2 protons and 2 neutrons (since its molar mass is 4.0 g and $4 - 2 = 2$). Similarly, a beryllium (**Be**) generally has 4 protons and 5 neutrons (since $9 - 4 = 5$). Look at the periodic table to see where the first two numbers in the equation are coming from (we will also discuss this more in the next section). The elements in the top few periods (rows) of the periodic table have approximately the same number of neutrons as they have protons. However, the elements in the bottom period have many more neutrons than protons. For example, a lead (**Pb**) atom has 82 protons and is expected to have about 125 neutrons (since $207 - 82 = 125$).

Why do the elements near the top of the periodic table have approximately the same number of neutrons as protons, while the elements near the bottom of the periodic table have many more neutrons than protons? There are two parts to this answer. First, atomic nuclei get larger going down the periodic table because there are many more protons (and hence neutrons, too) in the nucleus as you go down the periodic table. As the nuclei get larger, the more distant protons are further apart. As the protons get further apart, their electrical repulsion increases compared to their strong nuclear attraction. Therefore, the nucleus must have additional neutrons – electrically neutral particles that increase the strong nuclear attraction without increasing the electrical repulsion – in order to hold together.

Protons and neutrons (unlike electrons) are themselves composed of elementary particles called quarks. There are six different flavors of quarks: up, down, charm, strange, top, and bottom. The up, charm, and top quarks have a charge that is two-thirds the charge of a proton ($+2e/3$), while the down, strange, and bottom quarks have a charge that is one-third the charge of an electron ($-e/3$). The convention is to use e for the charge of a proton (not an electron), such that $-e$ is the charge of an electron (since it's negative). The value of e is 1.6×10^{-19} Coulombs (the SI unit of charge).

Although quarks are fractionally charged particles, quarks are never found in nature by themselves; they are always bound together in groups of two or three. All groups of quarks found in nature have integral (rather than fractional) charge. A proton consists of two up quarks and one down quark, so its charge is $\frac{2e}{3} + \frac{2e}{3} - \frac{e}{3} = \frac{3e}{3} = e$. A neutron consists of one up quark and two down quarks, so its charge is zero: $\frac{2e}{3} - \frac{e}{3} - \frac{e}{3} = 0$.

6.3 Isotopes

53
I
Iodine
126.9 g

magine that we have a pair of magic microscopic "pliers" with which we could add or remove protons, neutrons, or electrons to/from an atom. If you change the number of protons in the atom – with these "magic pliers" – you change the identity of the atom – i.e. you get a different element all together. For example, nitrogen (**N**) atoms always have 7 protons. If instead you change the number of neutrons, you get a different isotope of the atom. Lastly, if you change the number of electrons, you get an ion. For example, N^{3-} is an ion of nitrogen with 3 extra electrons.

Elements come in different isotopes – i.e. they do not always have the same number of neutrons in the nucleus. For example, a lead (**Pb**) atom with 128 neutrons is a different isotope than a lead atom that has 132 neutrons. A lead atom always has 82 protons, but lead atoms vary in their number of neutrons. We use a special notation – called a nuclide symbol – in order to distinguish between the different isotopes of an element. Before we introduce this notation, we will define atomic number and atomic mass number.

The atomic number, Z, equals the number of protons in the nucleus. The atomic number of each element can be found in the top right corner of each element's cell in the periodic table on pages 12-13. Each element has a unique atomic number.

The atomic mass number, A, equals the number of protons and neutrons in the nucleus. The atomic mass number approximately equals the mass of the atom because protons and neutrons have nearly the same mass and since electrons are very light compared to protons and neutrons. However, atomic mass is a little different from atomic mass number.

An element comes in a variety of isotopes. Isotopes differ in the number of neutrons in the nucleus. We use nuclide symbol notation to distinguish between the various isotopes. The nuclide symbol for a given

91 **32**
Pa **Ge**

 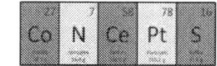

element, call it **X**, has the form $_Z^A\text{X}$, where **X** is the element's chemical symbol, Z is the atomic number (the number of protons), and A is the atomic mass number (the number of protons plus neutrons). For example, a thallium (**Tl**) atom with 125 neutrons has a nuclide symbol of $_{81}^{206}\text{Tl}$ because it has $Z = 81$ protons and $A = 81 + 125 = 206$ protons and neutrons. Compare this to $_{81}^{210}\text{Tl}$, which has 129 neutrons (since $210 - 81 = 129$).

atomic mass ≠ atomic mass number

m_X = **atomic mass** of element

- Atomic mass includes the mass of the protons, neutrons, electrons, and nuclear binding energy.

A = **atomic mass number** of element

- Atomic mass number only includes the protons and neutrons.

Atomic mass number only **approximates** the atomic mass.

These two terms are **not** interchangeable.

The **atomic mass** of Cesium (**Cs**) is **132.9 g**. This number appears on the periodic table.

The **atomic mass number** of $_{55}^{135}\text{Cs}$ is **135**. This is 80 + 55.

Atomic mass number is an integer. Atomic mass is a decimal.

 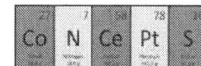

What is the atomic number of $^{235}_{92}U$?

The atomic number is:

$$Z = 92$$

It has 92 protons.

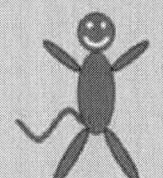

Which element is this?

What is the atomic mass number of $^{235}_{92}U$?

The atomic mass number is:

$$A = 235$$

How many neutrons are there? (143)

An iron isotope has 30 neutrons. Write its nuclide symbol.

The nuclide symbol is:

$$^{56}_{26}Fe$$

How many protons are there?

Given the nuclide symbol for an isotope, it is possible to determine the number of neutrons through subtraction. The atomic mass number, A, is the number of protons plus neutrons, while the atomic number, Z, is the number of protons. Therefore, the number of neutrons equals $A - Z$. For example, given the isotope $^{140}_{56}Ba$ of barium (Ba), subtract the atomic number ($Z = 56$) from the atomic mass number ($A = 140$) to determine that there are $A - Z = 140 - 56 = 84$ neutrons.

The isotopes of hydrogen (H) have special names. Protium ($^{1}_{1}H$) has no neutrons (it just has one proton and one electron), deuterium ($^{2}_{1}H$) has 1 neutron, and tritium ($^{3}_{1}H$) has 2 neutrons. The most common form of hydrogen is protium.

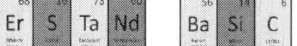 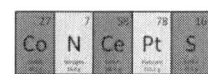
6.4 Radioactive Decays

 nstable nuclei can spontaneously decay into lighter nuclei by emitting an alpha (α), beta (β), or gamma (γ) particle. This process is called radioactive decay. The three nuclear decay channels are called alpha decay, beta decay, and gamma decay.

An alpha (α) particle is a helium (**He**) nucleus consisting of 2 protons and 2 neutrons. The nuclide symbol for an alpha particle is $_2^4\text{He}^{2+}$. An alpha particle is both a particular isotope of helium and an ion – it is a helium nucleus with no electrons. Therefore, an alpha particle has a positive charge of $+2e$ (two times the charge of one proton).

When an unstable nucleus decays via alpha decay, it loses 2 protons and 2 neutrons. Therefore, during alpha decay, the daughter nucleus has an atomic number that is 2 less than the parent nucleus and an atomic mass number that is 4 less than the parent nucleus (since $2 + 2 = 4$). In order to determine the daughter nucleus for alpha decay, find the element on the periodic table that has an atomic number that is 2 less than the parent nucleus. In order to write the nuclide symbol for the daughter nucleus in alpha decay, subtract 4 from the atomic mass number and subtract 2 from the atomic number. See the examples that follow.

A beta (β) particle is either an electron or the antiparticle of an electron – called a positron. A positron is just like an electron except for having positive charge; electrons and positrons have the same mass. Electrons are abundant in matter, but positrons are not. When a positron comes near an electron, the electron and positron annihilate, producing a pair of photons.[5] We will use the symbol $_{-1}^{0}\beta$ for an electron (the atomic mass number is 0 because there are no protons or neutrons, and the atomic number is -1 because it is negatively charged) and $_{+1}^{0}\beta$ for a positron. With the beta

---

[5] This shows that rest-mass is not conserved in all reactions that occur in nature (since the initial-state electron and positron both have rest-mass, but the final state photons – particles of light – do not), and so mass is not a fundamental conservation law like charge and energy.

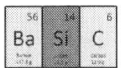 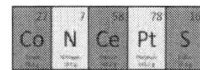

particle, the $+1$ or -1 in the place of the atomic number indicates the charge of the particle (not the number of protons).[6]

What is the product of the alpha decay of bismuth-210?

- The nuclide symbol for bismuth-210 is $^{210}_{83}\text{Bi}$.
- The product will have 2 fewer protons and 2 fewer neutrons. Therefore, the product is $^{206}_{81}\text{Tl}$. (Thallium)

$$^{210}_{83}\text{Bi} \rightarrow {}^{206}_{81}\text{Tl} + {}^{4}_{2}\text{He}$$

What is the product of the alpha decay of uranium-233?

- The nuclide symbol for uranium-233 is $^{233}_{92}\text{U}$.
- The product will have 2 fewer protons and 2 fewer neutrons. Therefore, the product is $^{229}_{90}\text{Th}$. (Thorium)

$$^{233}_{92}\text{U} \rightarrow {}^{229}_{90}\text{Th} + {}^{4}_{2}\text{He}$$

---

[6] Don't confuse the positron with the proton. Both have the positive charge of a proton, but one is an anti-electron. The positron has much less mass than a proton. Protons are found in nuclei, while positrons are not. Also, protons and neutrons are composed of quarks, but electrons and positrons are themselves elementary particles.

In beta emission, a neutron decays into a proton plus an electron: $^1_0n \rightarrow {}^1_1p + {}^0_{-1}\beta$. (Notice that both the atomic number and atomic mass number satisfy the reaction: That is, $1 = 1 + 0$ and $0 = 1 - 1$.) Carbon-14 decays to nitrogen-14 through beta emission, for example: $^{14}_6C \rightarrow {}^{14}_7N + {}^0_{-1}\beta$. In beta emission, the number of neutrons decreases and the number of protons increases. In order to determine the daughter nucleus for beta emission, add 1 to the atomic number (the atomic mass number remains unchanged).

An unstable nucleus can decay by emitting an alpha, beta, or gamma particle.

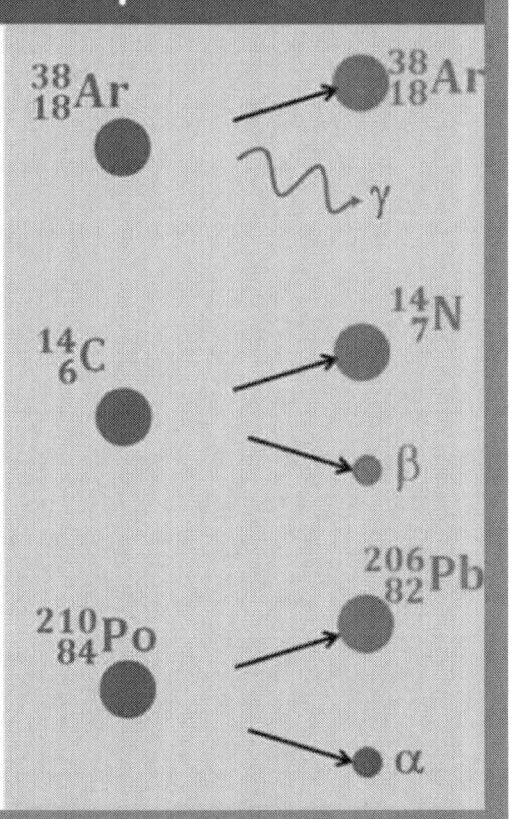

- A gamma particle (γ) is a photon – a particle of light. Gamma decay occurs when the same nucleus drops to a lower energy level.
$$^{38}_{18}Ar \rightarrow {}^{38}_{18}Ar + \gamma$$

- A beta particle (β) is an electron or positron (anti-electron).
$$^{14}_6C \rightarrow {}^{14}_7N + {}^0_{-1}\beta$$

- An alpha particle (α) is a helium nucleus (^4_2He) consisting of two protons and two neutrons.
$$^{210}_{84}Po \rightarrow {}^{206}_{82}Pb + {}^4_2He$$

In positron emission, the number of protons decreases by releasing a positron, which increases the number of neutrons: $^1_1p \rightarrow {}^1_0n + {}^0_{+1}\beta$. For example, bromine-80 decays into selenium-80 through positron emission:

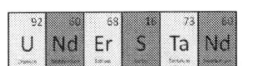 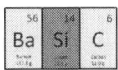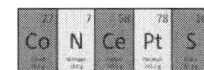
$^{80}_{35}\text{Br} \rightarrow\ ^{80}_{34}\text{Se} +\ ^{0}_{+1}\boldsymbol{\beta}$. In order to determine the daughter nucleus for positron emission, subtract 1 from the atomic number (the atomic mass number remains unchanged).

A gamma (**γ**) particle is a photon – a particle of electromagnetic radiation (which is a fancy name for light). A gamma particle has zero rest-mass (yet light is never at rest; photons do carry relativistic mass since they have energy according to Einstein's famous equation, $E = mc^2$). In gamma decay, an atom in an excited state emits a photon and drops down into a less excited state (such as the ground state). In the case of gamma decay, the parent and daughter nuclei are the same – the only difference is that the parent nucleus is a more excited state of the atom.

6.5 Half-Life

 alf-life is a measure of the rate at which an unstable radioactive nucleus decays. More precisely, the half-life of a substance is the time it takes for one-half of the substance to decay. The longer the half-life, the slower the rate of the decay.

For example, suppose that we start with 800 g of thorium-234 ($^{234}_{90}\text{Th}$). Every day, we measure the sample and find that there is less thorium-234. The reason that there is less thorium-234 is that it decays into protactinium-234 via beta emission: $^{234}_{90}\text{Th} \rightarrow\ ^{234}_{91}\text{Pa} +\ ^{0}_{-1}\boldsymbol{\beta}$. Every day, there is less thorium-234 and more protactinium-234 because thorium nuclei are decaying into protactinium nuclei.

After 24 days, all that remains of the original 800-g sample of thorium-234 is 400 g. Since it takes 24 days for one-half of the sample to decay, the half-life of thorium-234 is 24 days.

How much thorium-234 is left after 48 days (from the starting time)? The answer is not zero! Instead, the answer is 200 g. Every 24 days, one-half of the sample decays. If you can make a table like the one that follows, you can solve half-life problems without doing any math. ☺

Put the initial mass and 0 time in the first row. You must take care to avoid the common mistake of entering the half-life in the first row. Next, enter half the mass and the full half-life in the second row. In each successive row, cut the mass in half and add the half-life (add a half-life, don't double the half-life). The other common mistake is to double the half-life instead of adding it: Make sure that you cut the mass in half, but add (not double) the half-life each time.

800 g	0 days
400 g	24 days
200 g	48 days
100 g	72 days
50 g	96 days

According to the table above, initially there was 800 g of thorium-234. After 24 days, 400 g of thorium-234 remained; after 48 days, 200 g of thorium-234 was left; and only 50 g of thorium-234 remained after 96 days.

Carbon-14 decays to nitrogen-14 through beta emission with a half-life of 5715 years: $^{14}_{6}C \rightarrow ^{14}_{7}N + ^{0}_{-1}\beta$. Since carbon is necessary for life, comparing the ratio of carbon-14 to nitrogen-14 is useful for determining the age of the remains of living things. We call this radiocarbon dating – and more generally (i.e. not necessarily using carbon) it is referred to as radiometric dating.

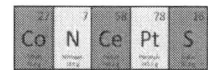

Cobalt-60 has a half-life of 10 minutes. If we start with a 80 g sample of cobalt-60, how much will remain 30 minutes later?

- In 10 minutes, 40 g remain.
- In 20 minutes, 20 g remain.
- In 30 minutes, **10 g remain.**

We start with a 48-g sample of phosphorus-32. Only 12 g remain 30 days later. What is the half-life of phosphorus-32?

- In x days, 24 g remain.
- In 2x days, 12 g remain. 2x = 30 days.
- Therefore, the half-life is **15 days**.

For example, if you find a substance that originally had carbon-14, but no nitrogen-14, and now it has 3 times as much nitrogen-14 as carbon-14, you can conclude that the substance is about 11,000 years old. The reason is as follows. One-half of carbon-14 decays into nitrogen-14 in 5715 years (the half-life of carbon-14). After one more half-life (another 5715 years, for 11,430 years total) three-fourths of the carbon-14 has decayed into nitrogen-14. The value of 3/4 comes from subtracting one-half of the one-half that remained; one-half of one-half is one-fourth, and one minus one-fourth is three-fourths. As an example, suppose that there was originally 12 g of carbon-14. After 5715 years, there is 6 g of carbon-14 and 6 g of nitrogen-14. After 11,430 years, there is 3 g of carbon-14 and 9 g of

nitrogen-14. There is 3 times as much nitrogen-14 as carbon-14 after about 11,000 years.

6.6 Nuclear Reactions

nlike chemical reactions, there is a change in the identity of the individual atoms involved in nuclear reactions. Consider, for example, the chemical reaction $Fe_2O_3 + 3CO \rightarrow 2Fe + 3CO_2$. In this chemical reaction, there is a change in chemical identity between the reactants, iron (III) oxide (Fe_2O_3) and carbon monoxide (CO), and the products, iron (Fe) and carbon dioxide (CO_2); but the same atoms – namely, iron (Fe), carbon (C), and oxygen (O) – are involved throughout the reaction. In contrast, consider the nuclear reaction $^{246}_{96}Cm + ^{12}_{6}C \rightarrow ^{254}_{102}No + 4^{1}_{0}n$. In this nuclear reaction, there is a change in identity between the initial atoms, curium (Cm) and carbon (C), and the final atoms, nobelium (No) and neutrons ($^{1}_{0}n$).

A very significant amount of energy may be released or absorbed in a nuclear reaction through Einstein's famous equation, $E = mc^2$, which expresses a correspondence between mass and energy, where $c = 300,000,000$ m/s is the speed that light travels in vacuum. According to this equation, the nuclear binding energy that holds protons and neutrons together in the nucleus of an atom is equivalent to an amount of mass given by $m = E/c^2$. Energy is conserved in all reactions – including nuclear and chemical reactions – and therefore conservation of energy is a fundamental law of nature. Rest-mass is not conserved in reactions, and so rest-mass does not follow a fundamental conservation law like energy and charge.

If the reactants have more rest-mass than the products, the reaction is exothermic as the difference in mass results in a release of energy according to $E = mc^2$. If instead the reactants have less rest-mass than the products, the reaction is endothermic and the reaction absorbs energy from the surroundings in order to conserve energy.

Einstein's famous equation: $E = mc^2$.

- This equation says that the **binding energy** that holds protons and neutrons together in the nucleus is equivalent to an amount of **mass** given by $m = E/c^2$.

- The isotope of helium, ^4_2He, contains 2 protons, 2 neutrons, and 2 electrons.

- But the atomic mass of ^4_2He differs from $2m_p + 2m_n + 2m_e$ by the **binding energy** that holds the nucleus together.

Radioactive isotopes have a lot of energy that can be harnessed through $E = mc^2$. The parent nucleus releases some of this energy when it decays to lighter components.

A heavy nucleus splits into two lighter nuclei, releasing an enormous amount of energy, in a nuclear reaction called nuclear fission. The parent nucleus is heavier than the combined mass of the daughter nuclei. The mass difference results in the release of energy through $E = mc^2$.

When one unstable radioactive nucleus decays through nuclear fission, one of the products can trigger a chain reaction. For this reason, nuclear fission has much practical application, and is used in nuclear reactors and atomic bombs.

In **nuclear fission,** a heavy nucleus splits into two lighter nuclei, releasing an enormous amount of energy.

- The parent nucleus is heavier than the combined mass of the daughter nuclei. The mass difference results in the release of energy: $E = mc^2$.

- When one unstable radioactive nucleus decays, one of the products can trigger a **chain reaction.**

- Nuclear fission is used in **nuclear reactors** and **atomic bombs.**

A chain reaction of nuclear fission

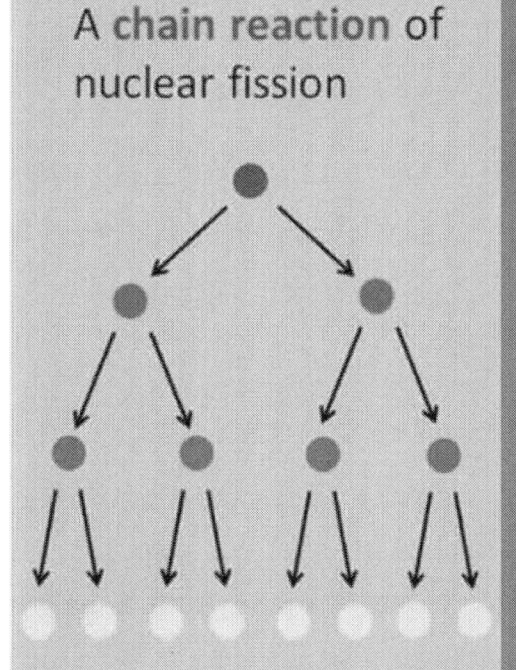

Two lighter nuclei combine to form a heavier, stable nucleus in a process called nuclear fusion (not to be confused with fission – fission involves splitting nuclei, whereas fusion involves combining nuclei together). The combined mass of the reactants exceeds the mass of the product. The mass difference releases even more energy than in nuclear fission.

Stars like our sun use nuclear fusion to make helium nuclei from hydrogen nuclei: $4^1_1\text{H} \rightarrow \,^4_2\text{He} + 2\,^0_{+1}\beta$. The vast amount of energy released

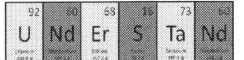 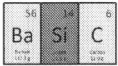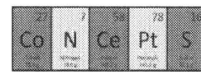
provides the energy that we need to sustain life on earth. Nuclear fusion is also used in hydrogen bombs.

Ch. 6 Self-Check Exercises

1. What is the atomic number of silicon (**Si**)?

2. Which element has an atomic number of 6?

3. If an isotope of calcium (**Ca**) has 22 neutrons, what is its atomic mass number?

4. How many protons are there in neon (**Ne**)?

5. If the atomic mass number of an isotope of iodine (**I**) is 127, how many neutrons does it have?

6. If the atomic mass number of an isotope of bismuth (**Bi**) is 209, how many neutrons does it have?

7. How many protons are there in $_{29}^{64}\textbf{Cu}$?

8. How many neutrons are there in $_{29}^{64}\textbf{Cu}$?

9. An isotope of an element has 50 protons and 70 neutrons. Write the nuclide symbol for this isotope.

10. Given the initial mass and half-life, determine the final mass at the specified time.

(A) $m_i = 16$ g, half-life = 4 days, find m_f 8 days later.

(B) $m_i = 88$ g, half-life = 3 hrs, find m_f 12 hrs later.

(C) $m_i = 256$ g, half-life = 30 min., find m_f 2 hrs later.

(D) $m_i = 112$ g, half-life = 5 yrs, find m_f 25 yrs later.

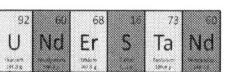 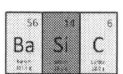 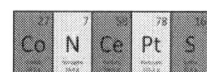
11. Given the initial mass and the final mass at the specified time, determine the half-life.

(A) $m_i = 32$ g, 15 min. later $m_f = 8$ g, find the half-life.

(B) $m_i = 160$ g, 8 wks later $m_f = 40$ g, find the half-life.

(C) $m_i = 400$ g, 60 yrs later $m_f = 25$ g, find the half-life.

(D) $m_i = 168$ g, 18 days later $m_f = 21$ g, find the half-life.

12. Given the radioactive isotope, write the nuclide symbol for the daughter nucleus that is formed when the specified parent nucleus undergoes alpha decay.

(A) Radium 222. Its symbol is **Ra**.

(B) Polonium 218. Its symbol is **Po**.

(C) Uranium 238. Its symbol is **U**.

(D) Thorium 230. Its symbol is **Th**.

(E) Astatine 218. Its symbol is **At**.

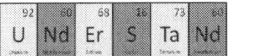 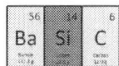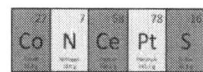
7 ☐ VErBAl ReAcTiONS

7.1 Chemical Words

 e conclude this text with a bit of a fun diversion – some verbal chemistry puzzles. You may have noticed several "chemical words" mixed in with the content of this book. A chemical word is a word that can be written exclusively by using symbols from the periodic table. An example of a chemical word is **ThInK**, which consists of the symbols for the following elements: thorium (**Th**), indium (**In**), and potassium (**K**).

Many words can't be made using just the symbols from the periodic table. Unfortunately – or perhaps ironically – the word "chemistry" can't be made this way: There is no way to make the 'm,' 't,' or 'r.' The only symbols on the periodic table with an 'm' are americium (**Am**), curium (**Cm**), fermium (**Fm**), magnesium (**Mg**), manganese (**Mn**), molybdenum (**Mo**), meitnerium (**Mt**), mendelevium (**Md**), promethium (**Pm**), samarium (**Sm**), and thulium (**Tm**), which makes it impossible to spell "chemistry" – since we would either need **M**, **Mi**, or **Em** to be a symbol on the periodic table in order to make the 'm.' Similarly, there doesn't exist a **T**, **R**, **St**, **Tr**, or **Ry** in order to make the 'tr' part of "chemistry."

There are many other common English words that can't be made using just symbols from the periodic table. For example, you can't make any words

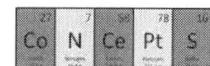

that have a 'j,' 'q,' or a 'z,' since these letters are not used in any of the symbols from the periodic table. Many words have beginnings or endings that can't be made just using chemical symbols – like words beginning with 'an' or ending with 'ing.' Many words also have common letter combinations like 'str' that preclude them from being possible chemical words.

Nonetheless, it is still possible to make thousands of chemical words that are common words from the English language, like **SUPErB**, **SeNSAtIONAl**, and **UNReAl**. It's rather amazing that so many words can actually be made using just the symbols from the periodic table – most of which come in pairs like **Nd**, **Li**, and **Cr**. Some of the symbols are not even useful for making words, like **Zr**, **Cf**, and **Bk**, and some of the most frequent letters of the English language only appear in pairs – that's the case with 'a,' 'e,' and 't,' for example.

Try staring at the periodic table on pages 12-13 for a while and see if you can come up with some good words. You might find that it's a refreshing verbal diversion from science.

One interesting feature of chemical words is the concept of "chemonyms." A "chemonym" is a word that can be spelled using more than one set of chemical symbols. For example, the word "because" can be spelled as **BeCaUSe** or **BeCAuSe**. One version uses calcium (**Ca**) and uranium (**U**), while the alternate spelling uses carbon (**C**) and gold (**Au**), to make the 'cau' part of "because."

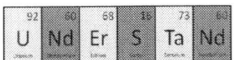 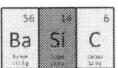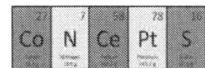
It is more challenging to try to construct chemical words where every symbol has two letters. The following chemical word is fascinating because it is composed of six two-letter symbols:

7.2 Chemical Word Puzzles

ne consequence of being able to make thousands of chemical words is that many common word puzzles – like crosswords, word scrambles, and word searches – can be made using just the symbols from the periodic table. It turns out that some word puzzles gain some interesting advantages by using chemical words rather than ordinary words.

For example, most popular word scramble and word jumble puzzle books use mostly four- to six-letter words, since it is much more difficult to rearrange the letters of longer scrambled words in order to think of the actual word. One advantage that chemical word scrambles have over ordinary word scrambles is that the letters of some of the symbols come in pairs. This allows chemical word scramble puzzles to use a vocabulary of longer words without increasing the difficulty of the puzzle. For example, the chemical word FAsCInAtEs has ten letters, but just six symbols. This means that unscrambling the six symbols CAtFInEsAs to form the chemical word FAsCInAtEs is comparable in difficulty to unscrambling the letters of an ordinary six-letter word, whereas it is much more difficult to unscramble the ten letters "catfinesas" to form the ordinary word "fascinates."

 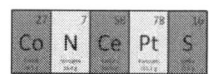

As a sample chemical word scramble, see if you can unscramble the following chemical symbols to form a single, six-symbol (nine-letter) common English word.

Spoiler alert: You can find the answer below if you keep reading. So if you haven't solved this chemical word scramble yet and don't want to spoil your fun, don't continue reading below until you have figured it out.

Consider that the above word would have been more difficult to solve if it consisted of nine individual letters as "unisogein" instead of six chemical symbols. The answer to the above word scramble is ingenious, ready or not.

If you enjoyed this chemical word scramble, you may want to try the chemical word scrambles in the exercises at the end of this chapter.

Chemical word searches are similarly more challenging than ordinary word searches. In part, this is because some of the cells have a two-letter chemical word symbol. An interesting feature of chemical word searches is that new chemical words can be formed by reading words backwards. For example, the word "ability" can't be made as a chemical word going forward, but you can write it backwards using the symbols **YTiLiBa** (of course, you must read this word right-to-left instead of left-to-right). Allowing chemical words to be written backwards in this way makes chemical word searches even more challenging. Note that some chemical words that can be written forwards can't be written backwards. For example, the chemical word

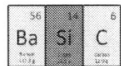 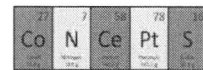
IrONiC can't be written in reverse. Some chemical words can be written both forwards and backwards, like **ReVErSe** and **EsReVEr**.

You can find a chemical word search in the exercises at the end of this chapter.

Crossword puzzles are much more challenging to make with chemical words than ordinary words – especially, if you try to make the fancy full-square, mirrored-about-the-diagonal crossword puzzles that are common in newspapers. There is a sample chemical crossword puzzle in the exercises at the end of this chapter.

The use of chemical symbols in composing cryptograms offers an added means of preventing outsiders from decoding a secret message: Someone trying to decode a message may not realize that some of the letters come in pairs. For example, suppose you want to code the message, "Attack the enemy at oh eight hundred," in such a way that someone who intercepts the message can't decode it. A simple thing that you might try is to change all of the a's into q's, the t's into e's, the c into a w, etc. In that case, the message might read as, "Qeeqwa ejo odopz qe kj obsje jcdfhof." The problem is that a basic cryptogram like this can easily be decoded by understanding the frequency of letters and common positioning of letters in the English (or other) language.

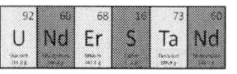 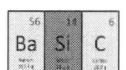
However, if some of the letters are exchanged for symbols from the periodic table, then the length of each word will change. Some of the original letters become a pair of letters instead of a single letter, while others become a different single letter. This adds an extra complication to ordinary cryptograms, as an unintended decoder doesn't know how long the words are and doesn't even know the correct positioning of many of the letters (since a letter that seems like it's the third letter could very well be a second letter). For example, if you change the a's into he's, the c into a w, the t's into es's, etc., the message might look like, "Heeseshewar esycn cnbcnpao hees hy cnxefyes ykrbhgnecnhg." You can easily introduce a lot of confusion by using the symbols **C**, **N**, and **Cn** to represent three different letters, for example.

You won't find a cryptogram with the end-of-chapter exercises since, as you can see, they can be very hard. We mention the chemical cryptogram for its own academic merit.

7.3 VErBAl ReAcTiONS

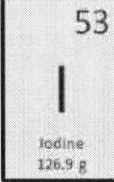

f you enjoy both chemistry and word puzzles, the perfect combination may be a chemical word puzzle called **VErBAl ReAcTiONS**. Even the name of the word puzzle uses chemical words!

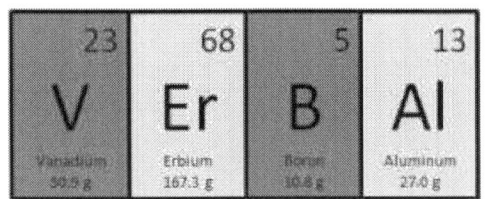

VErBAl ReAcTiONS are chemical word scrambles that are written in the form of chemical reactions. Just like chemical reactions, coefficients are

used to indicate when there are two or more of the same symbol involved in the chemical word scramble. For example, consider the **VErBAl ReAcTiON** below.

$$2\ \textbf{C} + \textbf{U} + 2\ \textbf{S} + \textbf{Es} \rightarrow\ \underline{\ \ }\ \underline{\ \ }\ \underline{\ \ }\ \underline{\ \ }\ \underline{\ \ }\ \underline{\ \ }$$

Both sides of this **VErBAl ReAcTiON** have 2 **C**'s, 1 **U**, 2 **S**'s, and 1 **Es**. The answer, which fits in the blanks on the right-hand side (one symbol per blank – so there could be two letters on the same blank, as in **Es**) has the same six symbols rearranged to form a common English word.

Spoiler alert: You can find the answer below if you keep reading. So if you haven't solved this **VErBAl ReAcTiON** yet and don't want to spoil your fun, don't continue reading onto the next paragraph until you have figured it out.

The chemical word success is the answer to the **VErBAl ReAcTiON** above.

If you enjoyed this **VErBAl ReAcTiON**, you may want to try the **VErBAl ReAcTiONS** in the exercises at the end of this chapter. You can also find a few **VErBAl ReAcTiONS** books available at your favorite online bookseller, such as Amazon or Barnes & Noble.

  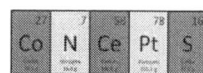

Ch. 7 Self-Check Exercises

1. Unscramble each set of chemical symbols to make a common English word.

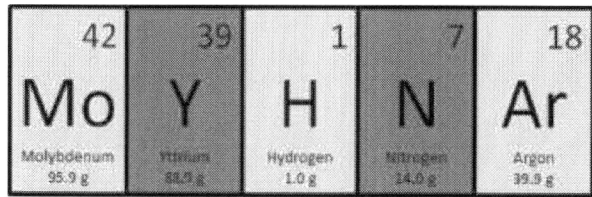

— — — — —

7	53	66	6	95
N	I	Dy	C	Am
Nitrogen	Iodine	Dysprosium	Carbon	Americium
14.0 g	126.9 g	162.5 g	12.0 g	243.1 g

— — — — —

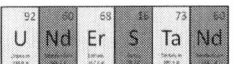 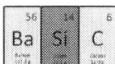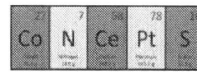

___ ___ ___ ___ ___ ___

___ ___ ___ ___ ___ ___

75	16	42	85	2	15
Re	S	Mo	At	He	P
Rhenium	Sulfur	Molybdenum	Astatine	Helium	Phosphorus
186.2 g	32.1 g	95.9 g	210.0 g	4.0 g	31.0 g

___ ___ ___ ___ ___ ___

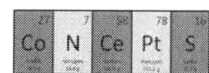

2. Find the chemical names listed below in the chemical word search puzzle on the next page. When chemical words read right-to-left (including right-to-left diagonals), read the symbols backwards (as explained in Sec. 7.2). Each of these chemical names can be written using only symbols from the periodic table.

Word List

Alberta	Brittany	Gretchen	Pauline
Alexis	Cameron	Harriet	Rebecca
Allison	Candice	Heather	Shannon
Annalisa	Caroline	Katherine	Valery
Barbara	Catherine	Lindsey	Vanessa
Bernice	Francesca	Marion	Veronica
Blanche	Francine	Nikki	Yvonne

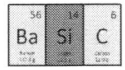 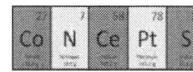
P	Re	S	I	Xe	La	V	F	N	V	Y	S	He	Ne	Am	At	I	Y	Xe	La
Au	B	Er	N	K	Re	Y	Ra	Re	Er	Al	Li	I	At	F	Nd	Re	V	Er	Rn
L	Ar	La	N	Ar	F	F	N	Rn	B	B	Er	Ta	K	Ca	Ra	Re	B	B	N
Rn	B	O	N	As	La	Ra	Be	Ti	Rb	H	Xe	Y	At	K	At	N	Ar	La	Ne
Re	C	Ar	O	Xe	Se	N	N	Am	At	S	V	La	V	Er	I	Li	B	B	Li
Te	Rg	At	Ba	Ra	O	Ce	Li	C	At	H	B	Ar	B	Ar	B	N	Ne	Re	Au
As	Ir	Ti	O	Ra	C	S	Nd	Ar	I	At	U	Na	V	Y	La	V	S	H	P
B	Se	Ra	Rb	N	H	Ca	R	O	N	Ne	Pa	U	L	K	N	Am	Nd	I	K
Er	Am	N	H	La	Er	R	Nd	Y	Ne	H	Ti	Rb	H	Ta	C	At	Li	Ra	At
N	Ir	N	Am	B	Ar	Ir	Ti	N	O	N	Na	H	S	Am	He	V	Er	He	H
N	Al	C	Ar	O	Li	Ne	P	Rb	Te	Rg	At	Ar	Ir	Ar	Y	Er	Al	Ra	Rn
I	Na	B	Er	F	Nd	Er	H	Au	Na	C	Am	Er	Re	B	Er	O	N	Li	Be
N	O	V	Y	N	S	Am	V	Ir	Te	Ir	Te	Al	Ar	Be	Am	N	K	He	S
O	Ca	Se	Er	V	Er	C	Am	Er	O	N	Rg	Rg	B	Ta	C	I	Ra	At	At
S	Nd	K	La	Al	O	P	Au	N	At	Te	At	Re	K	Na	Ar	Ca	F	H	Al
Li	I	Te	B	Er	V	N	P	Au	C	F	He	B	N	H	N	Nd	Ce	Er	Li
Al	I	Xe	Ir	Ar	Al	N	Ne	H	Ar	Ar	B	La	Ra	Al	Li	I	N	Ra	S
S	Nd	K	At	H	Er	I	Ne	N	O	S	I	Xe	Am	Ir	Rn	Ce	O	Na	H
V	C	Te	Rg	B	B	S	I	Xe	K	As	Se	Te	Ir	Be	Am	Er	Na	N	V
Y	N	At	Ti	Rb	Re	H	Ta	K	Be	Rn	As	Se	N	Na	V	Y	N	At	S

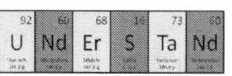

3. Solve the chemical crossword puzzle below.

Across

1. having uniform composition
6. ____'s exclusion principle
8. complex polymers found in cells; essential to animal diets
10. ____ rating (a measure of fuel efficiency)
12. prefix for one quadrillionth
13. occurs when a solvent passes through a membrane
15. H_2O
17. ionic ____
18. of opal
20. salts containing HCO_3^-
24. lab ____ (holders)
25. there are 3 ____ of matter
26. country
27. nona means ____
28. the stuff ____ life
33. basis for organic chemistry
34. C_9H_{20} alkane hydrocarbon
35. ____ materials (in their natural condition)
36. prefix for 20 across
37. a conductive metal
39. web____ (www)
41. ____ burner
43. short for ammunition
45. note of debt (abbrev.)
48. chemical ____
49. brief summary of a report
51. splitting atoms

52. a pro basketball association (abbrev.)
53. rope to prevent an animal from grazing
54. birds of a ____
55. a copper/zinc alloy
58. skin mark
60. British unit of pressure (abbrev.)
61. fine (abbrev.)
62. SI unit of pressure
64. salts containing $(PO_4)^{3-}$
66. ____ law, relating pressure/temperature (2 words, apostrophe)
69. formula for nitrogen monoxide
70. Swiss ____ (mountain range)
71. feminine pronoun
74. Dewar ____
76. soapy water with froth
77. forming a complex compound from simpler compounds
79. come in first place
81. iron (____) oxide (numerals)
83. coal oil
84. to connect or associate
85. crystalline ____
86. an atom's center

Down

1. a group of highly reactive nonmetals
2. prefix for one
3. singles
4. homogeneous mixture
5. ____ particle (an electron or positron)

7. carbon is the stuff of ____
8. C_3H_8 alkane used as fuel
9. basis for comparison among measurements
11. never used
14. matter of a particular chemical composition
16. wearing away
17. C_4H_{10} alkane used as fuel
19. fancy car
21. a positive ion
22. what the prefix deca means
23. to dissolve the maximum amount of solute in a solution
24. element that behaves like carbon
29. what instructors want their students to do during class
30. hereditary units
31. acid ____; weather condition
32. branch of physics dealing with heat transformations
33. electron ____ (a radioactive decay where a proton and electron produce a neutron)
36. substances that prevent the PH of a solution from changing
38. an oil spray; a woman's name
40. breakfast restaurant (abbrev.)

42. these prevent a container from leaking
44. ____ energy (minimum energy needed to transform the reactants)
45. ____ electrons (those nearest to the nucleus)
46. starch/water/alum/resin mixture used as an adhesive
47. toenail ____ (tools)
50. body parts offended by SO_2
55. ship
56. lawn tools
57. utensil with a broad, flat blade, used for spreading
59. an aluminum silicate with finely grained minerals; a form of soil
63. a limestone used for writing
64. force per unit area
65. where birds fly for the winter
66. ideal ___ (two words)
67. the court is now in ____
68. hydrochloric and sulfuric ____
72. ____ there! (interjection)
73. small bottles of fluids
74. includes gases and liquids
75. electron ____ (intrinsic angular momentum)
78. comment written in a journal
80. liquids used for writing
82. formula for hydrochloric acid

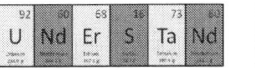 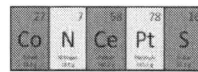
It may help to refer to the periodic table on pages 12-13 or on the back cover of this book. Place one symbol from the periodic table in each cell. Thus, there may be two letters in the same square.

4. Solve these VErBAl ReAcTiONS.

$$In + B + Ra + W + O \rightarrow \underline{}\ \underline{}\ \underline{}\ \underline{}\ \underline{}$$

$$Ar + C + 2\,He + Ta \rightarrow \underline{}\ \underline{}\ \underline{}\ \underline{}\ \underline{}$$

$$2\,C + P + 2\,I + N \rightarrow \underline{}\ \underline{}\ \underline{}\ \underline{}\ \underline{}\ \underline{}$$

$$Ar + H + 2\,P + Ra + Ag \rightarrow \underline{}\ \underline{}\ \underline{}\ \underline{}\ \underline{}\ \underline{}$$

$$2\,O + Ra + 2\,C + N \rightarrow \underline{}\ \underline{}\ \underline{}\ \underline{}\ \underline{}\ \underline{}$$

$$2\,Ti + Cs + 2\,S + Ta \rightarrow \underline{}\ \underline{}\ \underline{}\ \underline{}\ \underline{}\ \underline{}$$

$$C + H + I + 2\,S + P + Y \rightarrow \underline{}\ \underline{}\ \underline{}\ \underline{}\ \underline{}\ \underline{}\ \underline{}$$

$$C + At + I + 2\,Re + N + O \rightarrow \underline{}\ \underline{}\ \underline{}\ \underline{}\ \underline{}\ \underline{}\ \underline{}$$

$$2\,K + B + 4\,O + C \rightarrow \underline{}\ \underline{}\ \underline{}\ \underline{}\ \underline{}\ \underline{}\ \underline{}$$

$$2\,O + C + S + 2\,U + Ra + Ge \rightarrow \underline{}\ \underline{}\ \underline{}\ \underline{}\ \underline{}\ \underline{}\ \underline{}\ \underline{}$$

Re Answer Check

Ch. 1 Answers

1. chemical: milk spoils, hydrogen and oxygen combine to form water.
2. elements: lead, tungsten; compounds: salt, **MgO**; solution: milk; heterogeneous mixture: paper.
3. **He** (1,8), **Ba** (6,2), **I** (5,7), **Al** (3,3).
4. solids: **Os**, **I**; liquid: **Hg**; gas: **H**.
5. metal: calcium (**Ca**).
6. more active: **Ca**, **N**, **O**, **Fr**.
7. smaller: **Ti**, **Po**.
8. **N**, **K**, **Cs**, **Ir**.
9. halogen (**Br**), alkali (**Na**), noble gas (**Xe**), transition metal (**Au**).

Ch. 2 Answers

1. 2 (**He**), 12 (**Mg**), 29 (**Cu**), 92 (**U**), 0 (H^+), 19 (Ca^+), 24 (Fe^{2+}), 47 (Sn^{3+}), 86 (Fr^+), 2 (H^-), 10 (O^{2-}), 17 (S^-), 36 (Br^-), 54 (Te^{2-}).
2. 2 (**He**), 1 (**Na**), 2 (**Mg**), 3 (**Al**), 7 (**Cl**), 8 (K^+), 8 (Mg^{2+}), 1 (Al^{2+}), 18^7 (Ba^{2+}), 32^7 (Ra^+), 7 (O^-), 7 (N^{2-}), 8 (P^{3-}), 1 (Ar^-), 8 (Cl^-).
3. ionic: $MgCl_2$, Na_2O, Fe_2O_3.

---

[7] In a more advanced treatise, you would base your answer on the subshell structure. At any rate, the shell is full.

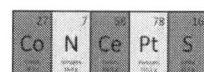

4. diatomic: H_2, Br_2.

5. Li^+, N^{3-}, Ba^{2+}, Se^{2-}, Sr^{2+}, F^-, O^{2-}, Be^{2+}, Al^{3+}, P^{3-}, Cl^-, Cs^+, S^{2-}, Rb^+.

6. LiF, $BeCl_2$, K_2S, Mg_3N_2, CaS, Al_2O_3, SrO, AlF_3, $FrCl$, Sr_3N_2.

7. H_2S; **S** shares 1 electron with each **H**.

Ch. 3 Answers

1. lithium fluoride, carbon dioxide, ammonia, francium oxide, iron (III) chloride, sodium hydroxide, copper (II) sulfate, barium chloride, dinitrogen trioxide, sodium chloride, tin (II) nitrate, sulfuric acid, phosphorus trichloride, magnesium fluoride, ammonium sulfite.

2. CO, H_2O, $CaCl_2$, $FeBr_2$, N_2O_4, Na_2O, BeF_2, $SnCl_2$, N_2O, $NaNO_3$, $(NH_4)_2O$, $Fe_2(SO_4)_3$.

Ch. 4 Answers

1. 4 (**Al**), 3 (**C**); 12 (**C**), 22 (**H**), 11 (**O**); 1 (**Mg**), 2 (**Cl**), 6 (**O**); 8 (**N**), 12 (**O**); 6 (**H**), 3 (**S**), 12 (**O**); 10 (**C**), 30 (**H**); 4 (**Sn**), 8 (**N**), 24 (**O**); 2 (**Ba**), 4 (**N**), 12 (**O**); 15 (**Hg**), 10 (**P**), 40 (**O**); 1 (**Al**), 3 (**N**), 9 (**O**); 2 (**N**), 8 (**H**), 2 (**Cr**), 7 (**O**); 16 (**K**), 8 (**C**), 24 (**O**); 18 (**H**), 6 (**P**), 24 (**O**); 8 (**C**), 32 (**H**), 8 (**N**); 6 (**N**), 24 (**H**), 3 (**S**), 9 (**O**).

 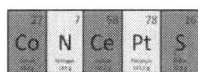

2. Balance the following chemical reactions:

$\mathbf{Sn \ + 2\,Cl_2 \ \rightarrow SnCl_4}$
$\mathbf{Br_2 \ + 3\,F_2 \ \rightarrow 2\,BrF_3}$
$\mathbf{4\,Al \ + 3\,O_2 \ \rightarrow 2\,Al_2O_3}$
$\mathbf{16\,Rb \ + \ S_8 \ \rightarrow 8\,Rb_2S}$

$\mathbf{2\,H_2S \ + 3\,O_2 \ \rightarrow 2\,H_2O \ + 2\,SO_2}$
$\mathbf{Al_2O_3 \ + 6\,HCl \rightarrow 2\,AlCl_3 \ + 3\,H_2O}$
$\mathbf{6\,Na \ + \ Fe_2O_3 \ \rightarrow 2\,Fe \ + 3\,Na_2O}$
$\mathbf{8\,Al \ + 3\,Fe_3O_4 \ \rightarrow 4\,Al_2O_3 \ + 9\,Fe}$

$\mathbf{3\,H_2S \ + 2\,HNO_3 \ \rightarrow 3\,S \ + 2\,NO \ + 4\,H_2O}$
$\mathbf{4\,NH_3 \ + 5\,O_2 \ \rightarrow 4\,NO \ + 6\,H_2O}$
$\mathbf{6\,CaO \ + \ P_4O_{10} \ \rightarrow 2\,Ca_3(PO_4)_2}$
$\mathbf{Al_2(SO_4)_3 \ + 3\,Ca(OH)_2 \ \rightarrow 2\,Al(OH)_3 \ + 3\,CaSO_4}$

Ch. 5 Answers

1. more dense: aluminum, ice, lead.
2. gently pressing a needle into it; pressure.
3. a 50-m diameter lake has a 100-m depth.
4. lead.
5. wood.
6. same.
7. Fahrenheit: 176°F, −100°F.

 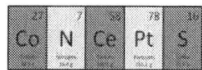

8. Celsius: 49°C, 127°C.

9. Kelvin: 283 K, 266 K.

10. 40 cc.

11. −123°C.

12. amorphous: rubber; covalent crystal: diamond; ionic crystal: sodium chloride; metallic crystal: aluminum; molecular crystal: dry ice

13. solute: sugar.

14. bases: oven cleaner, $Ba(OH)_2$. (Note that soda is cola, not baking soda.)

15. strong acid: pH of 2; weak acid: pH of 5; neutral: pH of 7; weak base: pH of 9; strong base: pH of 13.

Ch. 6 Answers

1. 14.

2. carbon.

3. 42. (It is not 40.1 g!)

4. 10.

5. 74.

6. 126.

7. 29.

8. 35.

9. $^{120}_{50}Sn$.

10. (A) 4 g (B) 5.5 g (C) 16 g (D) 3.5 g.

11. (A) 7.5 min. (B) 4 wks (C) 15 yrs (D) 6 days.

12. (A) $^{218}_{86}Rn$ (B) $^{214}_{82}Pb$ (C) $^{234}_{90}Th$ (D) $^{226}_{88}Ra$ (E) $^{214}_{83}Bi$.

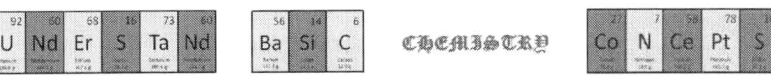

Ch. 7 Answers

1. HArMoNY, DyNAmIC, PaRaCHUTe, YArDsTiCK, AtMoSPHeRe

2.

P	Re	S	I	Xe	La	V	F	N	V	Y	S	He	Ne	Am	At	I	Y	Xe	La
Au	B	Er	N	K	Re	Y	Ra	Re	Er	Al	Li	I	At	F	Nd	Re	V	Er	Rn
L	Ar	La	N	Ar	F	F	N	Rn	B	B	Er	Ta	K	Ca	Ra	Re	B	B	N
Rn	B	O	N	As	La	Ra	Be	Ti	Rb	H	Xe	Y	At	K	At	N	Ar	La	Ne
Re	C	Ar	O	Xe	Se	N	N	Am	At	S	V	La	V	Er	I	Li	B	B	Li
Te	Rg	At	Ba	Ra	O	Ce	Li	C	At	H	B	Ar	B	Ar	B	N	Ne	Re	Au
As	Ir	Ti	O	Ra	C	S	Nd	Ar	I	At	U	Na	V	Y	La	V	S	H	P
B	Se	Ra	Rb	N	H	Ca	R	O	N	Ne	Pa	U	L	K	N	Am	Nd	I	K
Er	Am	N	H	La	Er	R	Nd	Y	Ne	H	Ti	Rb	H	Ta	C	At	Li	Ra	At
N	Ir	N	Am	B	Ar	Ir	Ti	N	O	N	Na	H	S	Am	He	V	Er	He	H
N	Al	C	Ar	O	Li	Ne	P	Rb	Te	Rg	At	Ar	Ir	Ar	Y	Er	Al	Ra	Rn
I	Na	B	Er	F	Nd	Er	H	Au	Na	C	Am	Er	Re	B	Er	O	N	Li	Be
N	O	V	Y	N	S	Am	V	Ir	Te	Ir	Te	Al	Ar	Be	Am	N	K	He	S
O	Ca	Se	Er	V	Er	C	Am	Er	O	N	Rg	Rg	B	Ta	C	I	Ra	At	At
S	Nd	K	La	Al	O	P	Au	N	At	Te	At	Re	K	Na	Ar	Ca	F	H	Al
Li	I	Te	B	Er	V	N	P	Au	C	F	He	B	N	H	N	Nd	Ce	Er	Li
Al	I	Xe	Ir	Ar	Al	N	Ne	H	Ar	Ar	B	La	Ra	Al	Li	I	N	Ra	S
S	Nd	K	At	H	Er	I	Ne	N	O	S	I	Xe	Am	Ir	Rn	Ce	O	Na	H
V	C	Te	Rg	B	B	S	I	Xe	K	As	Se	Te	Ir	Be	Am	Er	Na	N	V
Y	N	At	Ti	Rb	Re	H	Ta	K	Be	Rn	As	Se	N	Na	V	Y	N	At	S

3.

H	O	Mo	Ge	Ne	O	U	S		Be		P	Au	Li		Pr	O	Te	In	S
Al		N		Ne		O	C	Ta	Ne			Fe	Mt	O					Ta
O	Sm	O	S	I	S		Lu			W	At	Er			Pa		B	O	Nd
Ge		U			Ti						O	Pa	Li	Ne			U		Ar
N		B	I	Ca	Rb	O	Na	Te	S		Si		Mo		S	Ta	N	Ds	
S	Ta	Te	S		Ti		N		N	At	I	O	N		N	I	Ne		
		Ta		O	F		Ge		U		N		Ra		Li				Th
Ca	Rb	O	N		N	O	Na	Ne		Ra	W		B	I		Co	P	P	Er
Pt		Ce		C		S	I	Te		B	U	N	Se	N				Am	Mo
U		Ac		I	O	U		Ho		Pa		F		Al		Cl			Dy
Re	Ac	Ti	O	N		S	Y	No	P	Si	S		F	I	S	S	I	O	N
	V		N	Ba		Se		Te	Th	Er							P		Am
	Fe	At	H	Er		B	Ra	S	S		S	C	Ar				P	S	I
	I			O	K		P	As	C	Al		La			Er				C
P	H	O	S	P	H	At	Es		At		H		Ga	Y	Lu	S	S	Ac	S
Re	N	O			U		Al	P	S			Es						I	
S	He	U		V		F	La	S	K		La					S	U	Ds	
S	Y	N	Th	Es	I	S		Lu		P		W	In		I				
U		O		Al		I	I	I		H		K	Er	O	Se	Ne			
Re	La	Te		S	O	Li	Ds		N	U	Cl	Eu	S		N				

4. RaInBOW, HeArTaCHe, PICNIC, PArAgRaPH, RaCCOON, STaTiSTiCs, PHYSICS, ReCReAtION, COOKBOOK, COURaGeOUS.

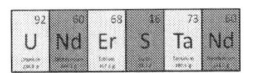 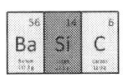 

About the Author

hris McMullen is a physics and astronomy instructor at Northwestern State University of Louisiana. He earned his Ph.D. in physics at Oklahoma State University in phenomenological high-energy physics (particle physics). His doctoral dissertation was on the collider phenomenology of superstring-inspired large extra dimensions, a field in which he has coauthored several papers.

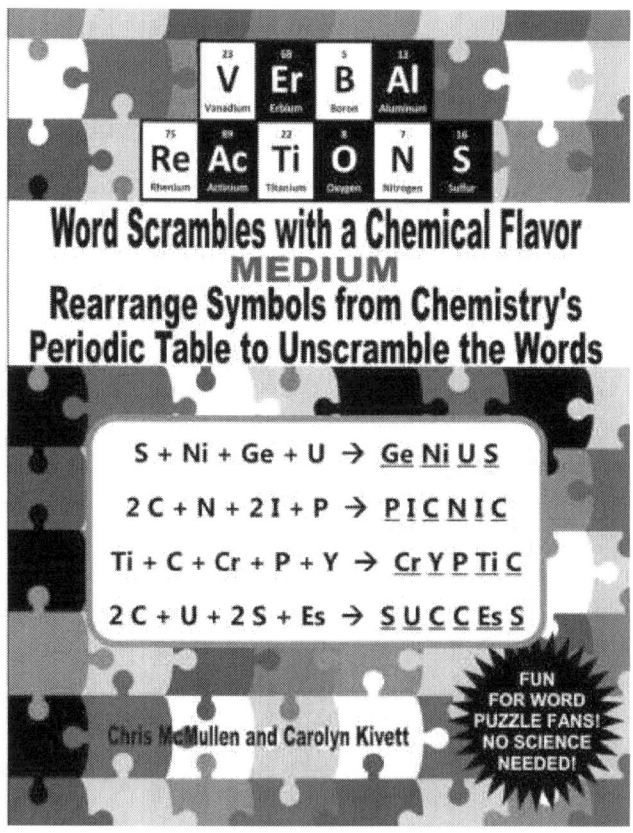

16261703R00094

Printed in Great Britain
by Amazon